AF753102

Antibacterial Peptide Protocols

METHODS IN MOLECULAR BIOLOGY™

John M. Walker, Series Editor

83. **Receptor Signal Transduction Protocols,** edited by *R. A. J. Challiss, 1997*
82. **Arabadopsis Protocols,** edited by *José M Martinez-Zapater and Julio Salinas, 1997*
81. **Plant Virology Protocols,** edited by *Gary D. Foster, 1997*
80. **Immunochemical Protocols (2nd. ed.),** edited by *John Pound, 1997*
79. **Polyamine Protocols,** edited by *David M. L. Morgan,* 1997
78. **Antibacterial Peptide Protocols,** edited by *William M. Shafer,* 1997
77. **Protein Synthesis:** *Methods and Protocols,* edited by *Robin Martin, 1997*
76. **Glycoanalysis Protocols,** edited by *Elizabeth F. Hounsel, 1997*
75. **Basic Cell Culture Protocols,** edited by *Jeffrey W. Pollard and John M. Walker, 1997*
74. **Ribozyme Protocols,** edited by *Philip C. Turner, 1997*
73. **Neuropeptide Protocols,** edited by *G. Brent Irvine and Carvell H. Williams, 1997*
72. **Neurotransmitter Methods,** edited by *Richard C. Rayne, 1997*
71. **PRINS and *In Situ* PCR Protocols,** edited by *John R. Gosden, 1997*
70. **Sequence Data Analysis Guidebook,** edited by *Simon R. Swindell, 1997*
69. **cDNA Library Protocols,** edited by *Ian G. Cowell and Caroline A. Austin, 1997*
68. **Gene Isolation and Mapping Protocols,** edited by *Jacqueline Boultwood, 1997*
67. **PCR Cloning Protocols:** *From Molecular Cloning to Genetic Engineering,* edited by *Bruce A. White, 1996*
66. **Epitope Mapping Protocols,** edited by *Glenn E. Morris, 1996*
65. **PCR Sequencing Protocols,** edited by *Ralph Rapley, 1996*
64. **Protein Sequencing Protocols,** edited by *Bryan J. Smith, 1996*
63. **Recombinant Proteins:** *Detection and Isolation Protocols,* edited by *Rocky S. Tuan, 1996*
62. **Recombinant Gene Expression Protocols,** edited by *Rocky S. Tuan, 1996*
61. **Protein and Peptide Analysis by Mass Spectrometry,** edited by *John R. Chapman, 1996*
60. **Protein NMR Protocols,** edited by *David G. Reid, 1996*
59. **Protein Purification Protocols,** edited by *Shawn Doonan, 1996*
58. **Basic DNA and RNA Protocols,** edited by *Adrian J. Harwood, 1996*
57. **In Vitro Mutagenesis Protocols,** edited by *Michael K. Trower, 1996*
56. **Crystallographic Methods and Protocols,** edited by *Christopher Jones, Barbara Mulloy, and Mark Sanderson, 1996*
55. **Plant Cell Electroporation and Electrofusion Protocols,** edited by *Jac A. Nickoloff, 1995*
54. **YAC Protocols,** edited by *David Markie, 1995*
53. **Yeast Protocols:** *Methods in Cell and Molecular Biology,* edited by *Ivor H. Evans, 1996*
52. **Capillary Electrophoresis:** *Principles, Instrumentation, and Applications,* edited by *Kevin D. Altria, 1996*
51. **Antibody Engineering Protocols,** edited by *Sudhir Paul, 1995*
50. **Species Diagnostics Protocols:** *PCR and Other Nucleic Acid Methods,* edited by *Justin P. Clapp, 1996*
49. **Plant Gene Transfer and Expression Protocols,** edited by *Heddwyn Jones, 1995*
48. **Animal Cell Electroporation and Electrofusion Protocols,** edited by *Jac A. Nickoloff, 1995*
47. **Electroporation Protocols for Microorganisms,** edited by *Jac A. Nickoloff, 1995*
46. **Diagnostic Bacteriology Protocols,** edited by *Jenny Howard and David M. Whitcombe, 1995*
45. **Monoclonal Antibody Protocols,** edited by *William C. Davis, 1995*
44. ***Agrobacterium* Protocols,** edited by *Kevan M. A. Gartland and Michael R. Davey, 1995*
43. **In Vitro Toxicity Testing Protocols,** edited by *Sheila O'Hare and Chris K. Atterwill, 1995*
42. **ELISA:** *Theory and Practice,* by *John R. Crowther, 1995*
41. **Signal Transduction Protocols,** edited by *David A. Kendall and Stephen J. Hill, 1995*
40. **Protein Stability and Folding:** *Theory and Practice,* edited by *Bret A. Shirley, 1995*
39. **Baculovirus Expression Protocols,** edited by *Christopher D. Richardson, 1995*
38. **Cryopreservation and Freeze-Drying Protocols,** edited by *John G. Day and Mark R. McLellan, 1995*
37. **In Vitro Transcription and Translation Protocols,** edited by *Martin J. Tymms, 1995*
36. **Peptide Analysis Protocols,** edited by *Ben M. Dunn and Michael W. Pennington, 1994*
35. **Peptide Synthesis Protocols,** edited by *Michael W. Pennington and Ben M. Dunn, 1994*
34. **Immunocytochemical Methods and Protocols,** edited by *Lorette C. Javois, 1994*
33. ***In Situ* Hybridization Protocols,** edited by *K. H. Andy Choo, 1994*
32. **Basic Protein and Peptide Protocols,** edited by *John M. Walker, 1994*
31. **Protocols for Gene Analysis,** edited by *Adrian J. Harwood, 1994*
30. **DNA–Protein Interactions,** edited by *G. Geoff Kneale, 1994*
29. **Chromosome Analysis Protocols,** edited by *John R. Gosden, 1994*
28. **Protocols for Nucleic Acid Analysis by Nonradioactive Probes,** edited by *Peter G. Isaac, 1994*
27. **Biomembrane Protocols:** *II. Architecture and Function,* edited by *John M. Graham and Joan A. Higgins, 1994*

METHODS IN MOLECULAR BIOLOGY™

Antibacterial Peptide Protocols

Edited by

William M. Shafer

Emory University School of Medicine, Atlanta, GA

Humana Press Totowa, New Jersey

© 1997 Humana Press Inc.
999 Riverview Drive, Suite 208
Totowa, New Jersey 07512

All rights reserved. No part of this book may be reproduced, stored in a retrieval system, or transmitted in any form or by any means, electronic, mechanical, photocopying, microfilming, recording, or otherwise without written permission from the Publisher. Methods in Molecular Biology™ is a trademark of The Humana Press Inc.

All authored papers, comments, opinions, conclusions, or recommendations are those of the author(s), and do not necessarily reflect the views of the publisher.

This publication is printed on acid-free paper. ∞
ANSI Z39.48-1984 (American Standards Institute) Permanence of Paper for Printed Library Materials.

Cover illustration: Fig. 1 from Chapter 10, "An Approach Combining Rapid cDNA Amplification and Chemical Synthesis for the Identification of Novel, Cathelicidin-Derived, Antimicrobial Peptides," by Alessandro Tossi, Marco Scocchi, Margherita Zanetti, Renato Gennaro, Paola Storici, and Domenico Romeo.

Cover design by Patricia F. Cleary.

For additional copies, pricing for bulk purchases, and/or information about other Humana titles, contact Humana at the above address or at any of the following numbers: Tel: 201-256-1699; Fax: 201-256-8341; E-mail: humana@mindspring.com, or visit our Website at www.humanapress.com

Photocopy Authorization Policy:
Authorization to photocopy items for internal or personal use, or the internal or personal use of specific clients, is granted by Humana Press Inc., provided that the base fee of US $8.00 per copy, plus US $00.25 per page, is paid directly to the Copyright Clearance Center at 222 Rosewood Drive, Danvers, MA 01923. For those organizations that have been granted a photocopy license from the CCC, a separate system of payment has been arranged and is acceptable to Humana Press Inc. The fee code for users of the Transactional Reporting Service is: [0-89603-408-9/97 $8.00 + $00.25].

Printed in the United States of America. 10 9 8 7 6 5 4 3 2 1

Library of Congress Cataloging in Publication Data

Main entry under title:

Methods in molecular biology™.

Antibacterial peptide protocols/edited by William M. Shafer.
p. cm.—(Methods in molecular biology; 78)
Includes bibliographical references and index.
ISBN 0-89603-408-9 (alk. paper)
1. Peptide antibiotics—Research—Methodology. 2. Peptide antibiotics—Analysis. I. Shafer, William M. II. Series: Methods in molecular biology (Totowa, NJ); 78.
RS431.P37A57 1997
615'.329—dc21 97-24698
CIP

Preface

As this book, *Antibacterial Peptide Protocols,* will attest, my enthusiasm for the field of antibacterial peptides is based on a conviction (and I am unashamed to say, prejudice) that these substances are in essence antibiotics produced by the host that then participate in host defense against infectious agents. Because of their capacity to exert antibiotic-like action against pathogenic microorganisms (bacteria, fungi, parasites, and viruses), there is reason to believe that these agents will soon be used clinically to treat infectious diseases. In fact, in recent years, biotechnology companies have been formed for the sole purpose of developing antibacterial peptides for clinical use. It should be emphasized that antibacterial peptides will likely play a major role in the treatment of infectious diseases, particularly with the increasing problem of multidrug-resistant microbes and the relative dearth of new antibiotics being provided by pharmaceutical companies.

The topic of this volume of *Methods in Molecular Biology,* the diverse methods used in research on antibacterial peptides, is thus quite timely. As the subject of antibacterial peptides develops into its own discipline (something strongly suggested by the explosion in the number of papers published over the past decade), it is essential that reliable techniques and strategies be made available not only to those of us in the field, but also to the newcomers and researchers in complementary disciplines. At first glance, the compilation of such techniques might seem an easy task, but, given the number and variety of techniques employed, it would be impossible for one person to take this on singlehandedly; after all, contemporary antimicrobial peptide research requires the availability of reliable protocols in microbiology, microbial genetics, cell biology, molecular biology, biochemistry, physical chemistry, in vivo experimentation, and histopathology. From the beginning, it was my purpose to assemble a team of experts who employ techniques in one or more of the aforementioned disciplines. Toward that end, I was extremely gratified when the numerous authors agreed to prepare chapters in their areas of expertise. Without exception, each author expressed a willingness to do so because of a general feeling that the field had reached a level of maturity that warranted the preparation of a methods-style handbook.

Antibacterial Peptide Protocols is divided into three main sections that are interconnected and preceded by a historical review of antibacterial peptides. The first section deals with the isolation and characterization of antimicrobial peptides. Herein, procedures involving the biochemical fraction of antimicrobial peptide-containing extracts from vertebrate and invertebrate sources and the physiochemical analysis of purified peptides are described in a series of chapters. Recent advances in the molecular characterization of genes encoding antimicrobial peptides, as well as procedures for using expression systems to isolate peptides, are described in the second section. Section three describes bioassays and microbial genetic techniques for studying antimicrobial peptides. It should be emphasized that the use of a reliable, standard bioassay for evaluating antimicrobial peptide activity is the necessary foundation for all studies dealing with this class of bioactive peptides. Toward this end, many different variations of antibacterial assays have been published and are used on a routine basis in different laboratories. The reader will notice this in chapters dedicated to other topics, but the authors include an antibacterial assay as the final readout system. An evaluation of the mechanisms by which peptides exert their antibacterial action are also covered in this section. Finally, it has become increasingly obvious in the past decade that these peptides possess activities other than their bactericidal effects. Thus, the capacities of certain peptides to inhibit the endotoxic effects of lipopolysaccharide and to serve as chemotactic agents for phagocytic cells have been documented, as is described in several chapters. These additional properties of antibacterial peptides may, in the final analysis, prove to be as important, if not more so, than their antibacterial action.

I trust that the reader will find each chapter of *Antibacterial Peptide Protocols* beneficial and helpful for his or her own particular needs. In closing, I wish to thank each author for their enthusiastic and diligent work in the preparation of this book. Special thanks are extended to Professor John Walker for his encouragement and advice, and to Mr. Thomas Lanigan for allowing me an extraordinary amount of freedom in choosing topics and authors.

William M. Shafer

Contents

Contributors

ARDEN ASPEDON • *Department of Biology, University of Missouri-St. Louis, St. Louis, MO*

MARY D. BARKLEY • *Department of Biochemistry, Louisiana State University, Baton Rouge, LA*

CHARLES L. BEVINS • *Research Institute, The Cleveland Clinic Foundation, Cleveland, OH*

DANIEL J. BRACKETT • *Departments of Surgery, Anesthesiology, and Pathology, University of Oklahoma Health Sciences Center, Department of Veterans Affairs Medical Center, Oklahoma City, OK*

PHILIPPE BULET • *Institut de Biologie Moléculaire et Cellulaire CNRS, Strasbourg Cedex France*

GILL DIAMOND • *Department of Anatomy, Cell Biology, and Injury Sciences, University of Medicine and Dentistry of New Jersey, Newark, NJ*

SUSAN W. FARMER • *Department of Microbiology and Immunology, University of British Columbia, Vancouver, British Columbia, Canada*

SHAFIQUE FIDAI • *Department of Microbiology and Immunology, University of British Columbia, Vancouver, British Columbia, Canada*

TOMAS GANZ • *Department of Medicine, University of California, Los Angeles, CA*

RENATO GENNARO • *Department of Biomedical Science and Technology, University of Udine, Italy*

EDUARDO A. GROISMAN • *Department of Molecular Microbiology, Washington University School of Medicine, St. Louis, MO*

ROBERT E. W. HANCOCK • *Department of Microbiology and Immunology, University of British Columbia, Vancouver, British Columbia, Canada*

CHARLES HETRU • *Institut de Biologie Moléculaire et Cellulaire CNRS, Strasbourg Cedex, France*

SADAAKI IWANAGA • *Department of Biology, Faculty of Science, Kyushu University, Fukuoka, Japan*

MARYAM M. JAVADPOUR • *Department of Biochemistry, Louisiana State University, Baton Rouge, LA*

MARTHA M. JUBAN • *Department of Biochemistry, Louisiana State University, Baton Rouge, LA*

SHUN-ICHIRO KAWABATA • *Department of Biology, Faculty of Science, Kyushu University, Fukuoka, Japan*
ROBERT I. LEHRER • *Department of Medicine, University of California, Los Angeles, CA*
MEGAN R. LERNER • *Departments of Surgery, Anesthesiology, and Pathology, University of Oklahoma Health Sciences Center, Department of Veterans Affairs Medical Center, Oklahoma City, OK*
KATHLEEN M. MORDEN • *Department of Biochemistry, Louisiana State University, Baton Rouge, LA*
TATSUSHI MUTA • *Department of Biology, Faculty of Science, Kyushu University, Fukuoka, Japan; Present Address: Department of Biochemistry, Kyushu University School of Medicine, Fukuoka, Japan*
H. ANNE PEREIRA • *Department of Pathology, University of Oklahoma Health Science Center, Oklahoma City, OK*
DOMENICO ROMEO • *Department of Biochemistry, Biophysics, and Macromolecular Chemistry, University of Trieste, Italy*
MARCO SCOCCHI • *Department of Biochemistry, Biophysics, and Macromolecular Chemistry, University of Trieste, Italy; National Laboratory of the Interuniversity Biotechnology Consortium, AREA Science Park, Trieste, Italy*
MICHAEL E. SELSTED • *Department of Pathology, College of Medicine, University of California, Irvine, CA*
JOHN K. SPITZNAGEL • *Department of Microbiology and Immunology, Emory University School of Medicine, Atlanta, GA*
DEBORAH A. STEINBERG • *IntraBiotics Pharmaceuticals, Inc., Sunnyvale, CA*
PAOLA STORICI • *Department of Biochemistry, Biophysics, and Macromolecular Chemistry, University of Trieste, Italy*
ALESSANDRO TOSSI • *Department of Biochemistry, Biophysics, and Macromolecular Chemistry, University of Trieste, Italy*
ERIKA V. VALORE • *Department of Medicine, University of California, Los Angeles, CA*
XIAOTANG WANG • *Department of Biochemistry, Louisiana State University, Baton Rouge, LA*
JERROLD WEISS • *Department of Microbiology, New York University School of Medicine, New York*
MARGHERITA ZANETTI • *National Laboratory of the Interuniversity Biotechnology Consortium, AREA Science Park, Trieste, Italy; Department of Biomedical Science and Technology, University of Udine, Italy*

1

Origins and Development of Peptide Antibiotic Research

From Extracts to Abstracts to Contracts

John K. Spitznagel

Peptide antibiotic research, which in the larger sense includes protein antibiotic research, actually began during the late 19th century with the work of Ehrlich, Metchnikov, Kanthack, and Petterson. Now it has been absorbed into the fields of microbiology, immunology, histochemistry, and cell biology. This early work depended on instruments, reagents, and techniques then at the cutting edge but now long since superseded: the compound microscopes, chemically characterized indicator stains, and the then-new science of bacteriology. Ehrlich, in 1879 defined the cytoplasmic granules of the granulocytic white blood cells. He noted that the granules of approx 2 or 3% of the cells stained intensely with eosin, an acid dye. He also noted that a much larger proportion of the granulated cells stained with eosin but also stained with the basic dye azur. Accordingly he designated the former cells eosinophils and the latter cells heterophils or neutrophils. He inferred from these staining properties that both kinds of cells carry basic proteins in their granules and that the neutrophil granules contain a mixture of basic and acidic proteins *(1)*. Metchnikov described in 1883 the preeminence of phagocytes including the neutrophils (microphages) in antimicrobial host defenses *(2)*. Kanthack and Hardy in 1895 discovered that phagocytosis of bacteria induced granulocytes to degranulate. They linked this degranulation with the death of the bacteria *(3)*. Petterson found antimicrobial activity in aqueous extracts of pus from human empyema; he attributed the action to basic proteins he found in the pus, comparing them to the protamines of salmon sperm *(4)*. Now, in retrospect, the necessary information might have been in place, at that time, to formulate a hypothesis con-

From: *Methods in Molecular Biology, Vol. 78: Antibacterial Peptide Protocols*
Edited by: W. M. Shafer, Humana Press Inc., Totowa, NJ

cerning the role of cationic granule proteins in host defenses against bacterial infection. As it happened, interest in the granules and their proteins had to lie fallow for more than 50 yr. The techniques of the time were simply unequal to the experimental demands.

Interest in the granules and their proteins rekindled as the era of cell biology opened and new methods for isolating cell organelles, separating cell proteins, purifying proteins, and characterizing proteins developed. The introduction of the lysosome concept by De Duve *(5)* profoundly influenced thinking about storage and delivery of antimicrobial and other discrete systems in phagocytes.

For example, a possible host defensive role for histone-like proteins aroused interest when Skarnes and Watson *(6)* reported that lactic acid extracts of rabbit polymorphs contained antimicrobial peptides with amino acid composition similar to histones and active against Gram-positive bacteria. They proposed that the histones of disintegrating polymorphs in pus would supply their nuclear histones to act as antimicrobial agents. The idea of the nucleus as a source of antimicrobial proteins understandably failed to invoke enthusiasm.

A more acceptable hypothesis based on the lysosome concept, soon developed that suggested the cytoplasmic granules of neutrophils as the storage organelles and delivery mechanism for antimicrobial substances. Thus the cytoplasmic granules of polymorphs received renewed interest owing to Robineaux and Frederic *(7)*, and Hirsch and Cohn *(8)*; the latter rediscovered degranulation with the help of the phase microscope. Hirsch and Cohn *(8)* demonstrated an antimicrobial activity extractable from polymorph granules with citric acid and dubbed it phagocytin. Phagocytin, like the leukins was antimicrobial in vitro. Were phagocytin and leukin the same things? Was phagocytin histone? Histones and the protamines were cationic proteins of the eukaryotic nucleus, bound electrostatically to DNA under physiological conditions. Intuitively, proteins bound to DNA seemed to be unlikely candidates for a major role in host defense; besides, DNA in vitro blocked the antimicrobial actions of histones *(9)*. Phagocytin, however, almost certainly a granule constituent, had a source and a delivery mechanism both plausible and suggestive. For a time it seemed possible that phagocytin was actually histone leached from the cell nuclei during preparation of the granule fraction and therefore really leukin. Moreover, the primary structures of both leukin and histones were unknown and it was not known whether cationic proteins other than histones existed in cells.

With histochemical methods, Spitznagel and Chi *(10)* showed that in guinea pig polymorphs the cytoplasmic granules stained strongly for very cationic arginine-rich proteins, and that when these cells phagocytized bacteria the granules aggregated around the bacteria and seemed to disappear. The cationic proteins then appeared to permeate the bacterial cells, rendering them histochemically positive for arginine-rich cationic proteins, substances that are foreign to bacteria. The killing of the bacteria correlated with the transfer to them of the cationic protein. These results, taken together with those of Hirsch and Cohn clearly pointed to the cytoplasmic granules of polymorphs as the sites of storage and the delivery mechanism for a heretofore undescribed antimicrobial cationic protein or proteins used by the phagocytes to kill bacteria. The question was whether the cationic protein(s) revealed by histochemistry were antimicrobial and whether they accounted for the death of the phagocytized bacteria.

Zeya and Spitznagel convincingly showed the existence of cationic antimicrobial protein-rich granules in neutrophils of guinea pigs, rabbits, and later of humans ***(11)***. This was done with differential centrifugation and paper electrophoresis that showed the granules of guinea pig polymorphs had indeed not one but several cationic antimicrobial proteins* (CAPS). They soon showed that the CAPs where present in other species as well. The proteins could be eluted from the paper and then freed of CTAB for antimicrobial assays. Interestingly, the experiment would not work without CETAB or some other cationic detergent. This suggested that the proteins were both cationic and hydrophobic.

The most cationic of these separated proteins were antimicrobial but showed no enzymic activity against substrates we tested, whereas other less cationic ones that did show enzyme activity were less antimicrobial. Electrophoretic studies failed to reveal any proteins with comparable cationic mobility or antimicrobial activity in extracts from cell nuclei removed from the cytoplasmic granules ***(12)***.

Amino acid analysis showed that the proteins had 25% arginine and 3.5% cysteine, features that clearly distinguished them from histones and are now considered characteristic of defensins (*see below*). They were rapidly bactericidal, inhibited the respiratory activity of *Eschericia coli*, and damaged bacterial permeability barriers. Bacterial cells irreversibly absorbed the proteins

---

*The present volume is concerned with techniques, and I feel it is worthwhile noting that Hirsch had attempted, unsuccessfully, to analyze phagocytin with starch block electrophoresis that time a state of the art electrophoretic technique (Hirsch, personal communication). H. I. Zeya, who had just joined me as a graduate student unsuccessfully tried something of the same kind with whole granules on paper electrophoresis. Zeya added cetyltrimethylammonium bromide (CETAB) to the buffer. It then occured to me that the setup was designed for the electrophoresis of serum proteins that have a range of isoelectric points (IEP) from 4–6.8. We were trying to separate proteins that our histochemistry had suggested might have IEP as high as 10 (Spitznagel and Chi). So, I had Zeya reverse the usual circuit by ignoring the instructions and attaching the positive power lead to the black binding post and the negative power lead to the red post. The result was that the proteins separated into several bands that moved to the negative pole.

from solution *(13)*. We called attention to the similarities between the antibacterial actions of the CAPs and those of polymyxin.

Zeya soon demonstrated with cationic sucrose density gradient electorphoresis that rabbit neutrophils have at least five cationic antibacterial proteins. The three most cationic proteins were nearly homogeneous and proved to have large arginine contents (34.7, 17.6, 6.6% of the total amino acids, respectively). Each had 14% cysteine. The arginine-rich protein fractions were different from each other both in amino acid composition and antimicrobial specificity. Gel filtration studies suggested that their size was less than 10 kDa. It was the first time that the antimicrobial specificities of the granule proteins and the chemical bases for their cationicity were made known. These highly cationic proteins were associated with the peroxidase-rich azurophil granules of rabbit polymorphs *(15)*.

Thus, in the early 1970s it was clear that the contents of neutrophil granules included heretofore unknown cationic peptides or proteins with antimicrobial action. The newly rediscovered phenomenon of degranulation provided these quintessentially phagocytic cells with an exquisitely precise and secure method for delivering these highly cytotoxic substances from the bone marrow to microbial invaders. Of course, these discoveries raised many questions about the biology and biochemistry of these substances and the mechanisms with which the phagocytic cells express them, store them in granules, and deliver them to target microbes. But, at that time the phenomenon seemed so simple that it was easy to dismiss the many challenging opportunities for careful investigation of these proteins and their actions in host defense.

In addition, in 1967, Holmes' discovery of the defect in polymorph, oxygen-dependent antimicrobial mechanisms in chronic granulomatous disease leukocytes *(16)* generated enormous interest in the pathophysiology of this X-linked oxidative killing defect. Her work plus the work of Klebanoff on the myeloperoxidase-H_2O_2-halide (MPO-H_2O_2-halide) killing system of neutrophils *(17)* thoroughly eclipsed interest in other polymorph antimicrobial mechanisms. This was not surprising considering that the basis for the defect in oxidative metabolism in chronic granulomatous disease phagocytes posed, in its own right, fascinating puzzles. Besides, Klebanoff promoted the apparently greater killing power (mol for mol) of the MPO-H_2O_2-halide system *(18)* compared to the granule cationic proteins. There is irony here since it was easily shown that birds do not have myeloperoxidase in their polymorphs *(19)* and later MPO deficiency in humans proved to have negligible effects on the health of the host *(20)*. Complete MPO deficiency occurs in about one in every 4000 people. The neutrophils of people with complete absence of MPO have reduced capacity to kill yeast, demonstrable in vitro. Klebanoff has described the biochemical characteristics of the deficient neutrophils in great detail *(21)*. It is striking that an enzyme like MPO, present in such large amounts in normal neutrophils and having such spectacular antimicrobial activity in vitro seems of so little consequence in host defense. (No doubt we are missing part of the equation!) In fact, it is noteworthy that very few clinically overt phenotypes have resulted from mutations in the granule proteins, probably owing to the redundancy of killing mechanisms.

There are several lines of evidence that the granule cationic proteins are important players in phagocytic host defenses:

1. They are carried in the azurophil granules *(22,23)*.
2. They are deposited into the phagolysosomes by degranulation, where they attach to and damage phagocytized particles *(24,25)*.
3. Phagocytized bacteria are killed in normal neutrophils under anaerobic conditions *(26)*.
4. Bacterial killing in chronic granulomatous disease (CGD) leukocytes is enhanced by bacterial H_2O_2 production *(27)*; however, certain bacteria (e.g., gonococci) not releasing H_2O_2 are killed by CGD leukocytes *(28)*.
5. Enteric bacteria exhibit endotoxin structure-dependent susceptibility to anaerobic neutrophils in a manner similar to their susceptibility to cationic proteins in vitro *(29)*.

At this point let's look at the development of knowledge of proteins derived from human and other mammalian nucleated blood cells. As we do so, we can examine the development of information concerning the antimicrobially active domains of these proteins. Then we will look at the discoveries of antimicrobial peptides in nonhematologic cells in mammals and other vertebrates. Finally, we will sketch out the discoveries of antimicrobial substances in insects, fish, and amphibian sources. Principal emphasis will be placed on the sources of these substances and the technics used to isolate and identify them.

In 1978 Weiss and Elsbach isolated a protein that they named BPI, bacterial permeability inducing factor, from a mass of granule proteins that had been accumulated over a period of 2 yr from neutrophils of a person with chronic myelogenous leukemia *(30)*. Gray et al. have cloned and sequenced the DNA that codes for BPI *(31)*. Shafer and colleagues independently discovered the BPI protein, which they designated CAP57 before they confirmed its homology with BPI. Shafer also described another antimicrobial protein, CAP37 *(32)* that has been confirmed by Gabay et al. who published the N-terminal 20 amino acids *(33)*. Pohl et al. revealed the complete amino acid sequence of CAP37 isolated from circulating mature neutrophils *(34)* and Morgan et al. cloned and sequenced the cdDNA that codes for CAP37 *(35)*. BPI/CAP57 and CAP37/Azurocidin are cationic and hydrophobic and have molecular weights of 57 and 37 kDa, respectively.

What is the nature of the other granule proteins and peptides? Lehrer and his colleagues, especially Selsted and Ganz, have demonstrated that low molecular weight species, that they have styled the defensins, comprise the bulk of cationic antimicrobial granule protein, approx 25%. First described in extracts from granules of rabbit peritoneal polymorphs by Zeya and Spitznagel ***(13,14)***, knowledge of the physical properties of the defensins was spectacularly extended by the work of Selsted ***(36,37)***. A number of other granule proteins with antimicrobial properties have been described. One, described by Holmes, who named it BP, seems to be identical with BPI/CAP57. All of these proteins have been confirmed by Scott ***(33)***. Many of the above proteins have been cloned and their amino acid sequences made known. Substantial structural details have been revealed for defensins and for CAP37 and CAP57/BPI. From the point of view of the present volume, the most exciting developments have occured as the structural basis for the antimicrobial action of these proteins and peptides have been solved. For example, Selstedt and colleagues have shown that defensins have three sulfhydryl bonds due to six highly conserved cysteines in each defensin molecule. They have also shown that these bands are essential for defensin antimicrobial activity ***(37,38)***.

Pereira and her colleagues have established that two nonhomologous domains of CAP37 are responsible, one for its antimicrobial and endotoxin binding actions ***(39)*** and the other for its chemotactic actions. Shafer et al. have shown that synthetic peptides based on the primary structure of cathepsin G are antimicrobial ***(40)***. They also have shown the importance of the guanidinium side chain of arginine in determining the bactericidal capacity of the cathepsin G-derived peptides ***(41)***. Interestingly, similar sequences sythesized with D amino acids have equal antimicrobial activity ***(40,41)***. With BPI, Ooi and her coworkers found that the N-terminal 24-kDa fragment of the 60-kDa holoprotein accounts for all of the antimicrobial and endotoxin binding actions ***(42)*** of the 60-kDa holo-BPI.

Scocchi et al. have described two antimicrobial proteins that they call bactenicins, Bac7 and Bac5 in extracts of bovine polymorph granules. These are proline- and arginine-rich polypeptides. In addition they have found that the bovine granules have a protein with 87% homology with CAP37 ***(43)***. This latter finding confirms unpublished observations showing CAP37 is immunohistochemically demonstrable in bovine neutrophil granules (Pereira, personal communication). Flodgard also has reported that a protein highly homologous with CAP37 can be isolated from porcine spleen. Moreover, he has solved its covalent structure ***(43)***. This long list of antimicrobial peptides now includes the cathelicidins ***(45)***, indolicidin ***(46)***, and ß-defensins ***(47)*** as well as other peptides that are described in subsequent chapters of this volume. One of the questions that has to be answered is whether all these peptides are primarily intended for antimicrobial action in host defense. Some have already been shown to have other actions. Both BPI and CAP37 bind and neutralize endotoxins. This may be intrinsically related to their antimicrobial action. CAP37, however, is a potent chemotaxin for monocytes, macrophages, and fibroblasts ***(44,48)***; defensins are also reported to have some chemotactic action ***(49)*** and certain defensins have corticostatic action ***(50)***. It is believable that some of these peptides have very important functions that remain to be recognized.

Cationic antimicrobial peptides also provide host defense in cold-blooded vertebrates. The serendipitous discovery of magainins in 1987 ***(51)*** by Zasloff and the work of Simmaco on bombinin ***(52)*** introduced an entirely new set of antimicrobial peptides that are proving of possible clinical therapeutic interest.

The field has also been greatly extended by the inclusion of peptides from invertebrate sources. Moreover, homologs of the insect peptides exist in vertebrates ***(53)*** and suggest the evolutionary importance of the antimicrobial peptides. The insect peptides were recognized as early as 1980 ***(54,55)***. As previously noted, homology has been demonstated between some mammalian peptides, the cryptidins, and the insect cecropins ***(56–58)***.

Other invertebrates express inducible antimicrobial peptides. In the horseshoe crab, *Limulus polyphemus*, infectious agents induce the release of antimicrobial substances, the tachyplesins, that are stored and carried in granules of this animal's hemocytes ***(59)***. Iwanaga and his colleagues have published a series of elegent papers characterizing the tachyplesins, their structure and mode of action. As with mammalian antimicrobial proteins such as BPI/CAP57, CAP37/azuricidin, cathepsin G, and the defensin peptides as well as the cecropins and the magainins, cationicity and amphilicity appear to be central to their antimicrobial properties. Important too are the formation of ß-sheet structures and the presence of cysteines and sulfhydryl bridges.

Overall the results with invertebrate antimicrobial peptides have been valuable not only because they provide new understanding of the molecular structures necessary for peptide antimicrobial action, but also because they show how widely the oxygen-independent defenses are distributed in the animal kingdom. In addition, the results show that these peptides tend to be located in sites apt to be in contact with our microbe-laden environments. Perhaps most significant, they are inducible in many settings. These facts add greatly to the conviction that the cationic antimicrobial peptides possess enormous survival benefits. Perhaps from the phylogenetic perspective they have been more important than oxidative killing mechanisms. Experience with insect peptides supports this concept. Several cationic antimicrobial peptides appear to have host defense roles in insects: apidaecins ***(60,61)*** and hymenoptaecin ***(62)***.

Still other antimicrobial peptides are reported from mammaliam sources: protegrins ***(63)*** and histatins ***(64)***. The great burst of activity that has added so many novel proteins and peptides to the list of antimicrobial peptides and proteins has stimulated investigators to extend earlier studies on their mode of action. These investigations have confirmed that net positive charge and amphilicity are characteristic of most of such molecules ***(66)***. Whether these features fully account for their activity are debatable, but they seem necessary in most instances and their positive charges are consistent with the capacity of the peptides to bind to microbial membranes bearing negative charges (*see below*). Their amphilicity is consistent with their capacity to damage cells by intercalating into hydophobic domains of their membranes.

Lehrer and his colleagues have shown that defensins form voltage-dependent channels in model membranes, which could explain their capacity to damage permeability barriers and to cause lysis ***(67)***. Whether they actually form complexes and lethal ion channels in microbial membranes remains to be seen ***(66)***. Lehrer and colleagues have also reported experiments to show that the defensins attack both the outer and the inner membranes of Gram-negative bacteria ***(68)***, and Weiss et al. have shown that BPI/CAP57 stops the respiratory activity of inverted inner membrane vesicles ***(69)***, a finding that recalls the early discovery by Zeya that the cationic proteins inhibit microbial respiration ***(13)***. Rest found that the granule antimicrobial proteins were most effective against log phase rough bacteria ***(69)***. This has been found to be the case by many investigators as they have worked with proteins purified from granule extracts.

With BPI, CAP37, and the pmrA mutant, *Salmonella typhimurium*, Shafer and, later, Roland have shown that the increased degree of 4-amino-arabinosylation of the phosphates on lipid A correlates with their increased resistance to the antimicrobial action of the proteins ***(70,71)***. This is consistent with the concept that, in order to initiate killing, the cationic proteins and peptides must react electrostatically with unsubstituted, negatively charged phosphates. Farley has shown that the sensitivity of *Salmonella* to killing by BPI is directly proportional to the binding of BPI to the bacterial cells. This binding is saturable and dependent on both positively charged and hydrophobic amino acids in the protein or peptide of interest ***(72)***. Groisman reports that *Salmonella* must have resistance to various host antimicrobial peptides in order to maintain virulence for mice ***(73)***. Roland finds that the *pmrA* locus defines a two-component regulatory system that along with *pmrD* in multiple copies determines resistance to polymyxin, cationic peptides, and proteins. The effects of these systems on host defense have not been determined ***(71,74)***. Other worthwhile work has been done with the various antimicrobial peptides as they have been identified in various species. Unfortunately, still more evidence is needed before the mechanisms of killing are clarified.

The articles presented in this volume will deal with the more recently described cationic antimicrobial peptides and protein in detail, so I have mentioned them only briefly. In addition, the most recent work on structure and function of peptides and proteins will be presented in detail. I hope that this introduction provides an historical context within which these developments can be appreciated. For many decades the principal motivation in the field was scholarly. In the past decade, however, and in step with the general commercialization of bioscience, prospects of application of these substances in clinical infectious disease have become a significant driving force toward development and in many respects a diversion. Thus, contemporoary investigators have pressed hard to discover and patent new molecules and to reveal the structural basis of antimicrobial action to provide bases for designing new synthetic or semisynthetic products with useful antimicrobial activities. In the succeeding chapters are many new answers and many new questions that have emerged from their efforts.

References

1. Ehrlich, P. and Lazarus, A. (1900) *Histology of the Blood.* (Myers, W., ed. and transl.) Cambridge University Press, Cambridge, UK. Reprinted in *The Collected Papers of Paul Ehrlich (1956) vol.1.Histology, biochemistry and pathology.* (Himmelweit, F., ed.) Pergamon Press, New York, pp. 181–268.
2. Metchnikov, E. (1905) *Immunity in Infective Disease.* (Binnie, F.G., transl.) Cambridge University Press, London, p. 198 ff.
3. Kanthack, A. A. and Hardy, W. B. (1895) The morphology and distribution of wandering cells of mammalia. *J. Physiol.* (Lond) 17;81.
4. Petterson, A. (1905) Ueber die bakteriziden leukocytenstoffe und ihre Beziehung zur Immuninitat. Centr. Bakteriol. *Parsitenk. Abt.* I **39**, 423–437.
5. De Duve, C. and Baudhuin, P. (1966). Peroxisomes (microbodies) and related particles. *Physio. Rev.* **46,** 323–357.
6. Skarnes, R.C. and D.W. Watson (1956) Characterization of leukin: an antibacterial factor from leucocytes active against gram-positive pathogens. *J. Exp. Med.* **104,** 829–45.
7. Robineaux, J. and Frederic, J. (1955) Contribution a l'etude des granulations neutrophiles des polynucleaires par la microcinematographie en contraste de phase. Compte Rendus des Seances de la Societe de Biologie (Paris). **149,** 486–489.
8. Hirsch, J. G. and Cohn, Z. A. (1960) Degranulation of polymorphonclear leucocytes following phagocytosis of microorganisms. *J. Exp. Med.* **112,** 105–114.

9. Spitznagel, J. K. (1961) Antibacterial effects associated with changes in bacterial cytology produced by cationic polypeptides. *J. Exp. Med.* **114,** 1079–1091.
10. Spitznagel, J. K. and Chi, H. Y. (1963) Cationic proteins and antibacterial properties of infected tissues and leukocytes. *Am. J. Pathol.* **43,** 697–711.
11. Zeya, H. I. and Spitznagel, J. K. (1963) Antibacterial and enzymic basic proteins from leukocyte lysosomes: separation and identification. *Science* **142,** 1085–1087.
12. Zeya, H. I. and Spitznagel, J. K. (1966) Cationic Proteins of polymorphonuclear leukocyte lysosomes: resolution of antibacterial and enzymatic activities. *J. Bact.* **91,** 750–754.
13. Zeya, H. I. and Spitznagel, J. K. (1966) Cationic proteins of polymorphonuclear leukocyte lysosomes II. Composition, properties, and mechanism of antibacterial action. *J. Bact.* **91,** 755–762.
14. Zeya, H. I. and Spitznagel, J. K. (1968) Arginine-rich proteins of polymorphonuclear leukocyte lysosomes. *J. Exp. Med.* **127,** 927–941.
15. Zeya, H. I. and Spitznagel, J. K. (1969) Cationic protein-bearing granules of polymorphonuclear leukocytes: separation from enzyme-rich granules. *Science* **163,** 1069–1071.
16. Holmes, B., Page, A. R., and Good, R. A. (1967) Studies of the metabolic activity of leukocytes from patients with a genetic abnormality of phagocytic function. *J. Clin. Invest.* **46,** 1422–1432.
17. Klebanoff, S. J. (1967) Iodination of bacteria: a bactericidal mechanism. *J. Exp. Med.* **126,** 1063–1078.
18. Klebanoff, S. J. and Clark, R. A. (1978) *The Polymorph: Function and Clinical Disorders.* North-Holland Press, Amsterdam, p. 458.
19. Brune, K. and Spitznagel, J. K. (1973) Peoxidaseless chicken leukocytes: isolation and characteriaction of antibacterial granules. *J. Infect. Dis.* **127,** 84–94.
20. Parry, M. F., Root, R. K., Metcalf, J. A., Delaney, K. K., Kaplow L. S., and Richar, W. J. (1981) Myeloperoxidase deficiency. Prevalence and clinical significance. *Ann. Int. Med.* **95,** 293–301.
21. Klebanoff, S. J. (1992) Oxygen metabolites from phagocytes, in *Inflammation: Basic Principles and Clinical Correlates, 2nd ed.* (Gallin, J. I. et al., eds.) Raven, New York, pp. 555–557.
22. Zeya, H. I. and Spitznagel, J. K. (1969) Cationic protein-bearing granules of polymorphonuclear leukocytes: separation from enzyme-rich granules. *Science.* **163,** 1069–1071.
23. Zeya, H. I. and Spitznagel, J. K. (1971) Characterization of cationic protein-bearing granules of polymorphonuclear leukocytes. *Lab. Invest.* **24,** 229–236.
24. Macrae, E. K. and Spitznagel, J. K. (1975) Ultrastructural localization of Cationic proteins in cytoplasmic granules of chicken and rabbit polymorphonuclear leukocytes. *J. Cell Sci.* **17,** 79–94.
25. Pereira, H. A., Spitznagel, J. K., Winton, E. F., Shafer, W. M., Martin, L. E., Gusman, G. S., Pohl, J., Scott, R. W., Marra, M. N., and Kinkade, J. M. (1980) The ontogeny of a 57-kD cationic antimicrobial protein of human polymorphonuclear leukocytes: location to a novel granule population. *Blood* **76,** 825–834.

26. Mandell, G. L. (1974) Bactericidal activity of aerobic and anaerobic polymorphonuclear neutrophils. *Infect. Immun.* **4,** 337–341.
27. Mandell, G. L. and Hook, E. W. (1969) Leukocyte bactericidal activity in chronic granulomatous disease: correlation of bacterial hydrogen peroxide production and susceptibillity to intracellular killing. *J. Bact.* **100,** 531–532.
28. Rest, R. F., Fischer, S. H., Ingham, Z. Z., and Jones, J. F. (1982) Interactions of *Neisseria gonorrhoeae* with human neutrophils: effect of serum and gonococcal opacity on phagocyte killing and chemiluminescence. *Infect. Immun.* **36,** 737–730.
29. Okamura, N. and Spitznagel, J. K. (1982) Outer membrane mutants of *Salmonella typhimurium* LT2 have lipopolysaccharide-dependent resistance to the bactericidal activity of anaerobic human neutrophils. *Infect. Immun.* **36,** 1086–1095.
30. Weiss, J., Elsbach, P., Olsson, I. and Odeberg, (1978) Purification and characterization of a potent bactericidal and membrane active protein from the granules of human polymorphonuclear leukocytes. *J. Biol. Chem.* **253,** 2664–2672.
31. Gray, P. W. G., Flaggs, G., Leong, S. R., Gumina, R. J., Weiss, J., Ooi, C. E., and Elsbach, P. (1989) Cloning of the cDNA of a human neutrophil bactericidal protein. Structural and functional correlation. *J. Biol. Chem.* **264,** 9505–9509.
32. Shafer, W. M., Martin, L. E., and Spitznagel, J. K. (1984) Cationic antimicrobial proteins isolated from human neutrophil granulocytes in the presence of diisopropyl fluorophosphate. *Infect. Immun.* **45,** 29–35.
33. Gabay, J. E., Scott, R. W., Campanelli, D. D., Griffith, J., Wilde C., Marra, M. N., Seeger, M., and Nathan, C. F. (1989) Antibiotic proteins of human polymorphonuclear leukocytes. *Proc. Natl. Acad. Sci. USA* **86,** 5610–5614.
34. Pohl, J., Pereira, H. A., Martin, M. N., and Spitznagel, J. K. (1990) Amino acid sequence of CAP37, a human neutrophil granule- derived antibacterial and monocytic-specific chemotactic glycoprotein structurally similar to neutrophil elastase. *FEBS Lett.* **272,** 200–204.
35. Morgan, J. F., Sukiennicki, T., Pereira, H. A., and Spitznagel, J. K. (1991) Cloning of the cDNA for the serine protease-like CAP37/Azurocidin, a microbicidal and chemotactic protein from human granulocytes. *J. Immunol.* **147**.
36. Selsted, M. E., Brown, D. M., DeLange, R. J., Harwig, S. S. L., and Lehrer, R. I. (1985) Primary structures of six antimicrobial peptides of rabbit peritoneal neutrophils. *J. Biol. Chem.* **260,** 4579–4584.
37. Selsted, M. E., Harwig, S. S. L., Ganz, T., Schilling, J. W., and Lehrer, R. I. (1985) Primary structures of three human neutrophil defensins. *J. Clin. Invest.* **76,** 1436–1439.
38. Selsted, M. E. and Harwig, S. S. (1989) Determination of the disulfide array in the human defensin HNP-2. A covalently cyclized peptide. *J. Biol. Chem.* **264,** 4003–4007.
39. Pereira, H. A., Erdem, I., Pohl, J., and Spitznagel, J. K. (1993) Synthetic bactericidal peptide based on CAP37: a 37kDa human neutrophil granule-associated cationic antimicrobial protein chemotactic for monocytes. *Proc. Natl. Acad. Sci. USA* **90,** 4733–4737.

40. Shafer, W. M., Shepherd, M. E., Boltin, B., Wells, L., and Pohl, J. (1993) Synthetic peptides of human lysosomal cathepsin G with antipseudomonal activity. *Infect. Immun.* **61,** 1900–1908.
41. Shafer, W. M., Hubalek, F., Huang, M., and Pohl, J. (1996) Bactericidal activity of a synthetic peptide (CG 117–136) of human lysosomal cathepsin G is dependent on arginine content. *Infect. Imm.* **64,** 482–485.
42. Ooi, C. E., Weiss, J., Elsbach, P., Frangione, B., and Mannion, B. (1987) A 25-kDa NH2-terminal fragment carries all the antibacterial activities of the human neutrophil 60-kDa bactericidal/permeability increasing protein. *J. Biol. Chem.* **262,** 14,891–14,894.
43. Scocchi, M., Romeo, D., and Zanetti, M. (1994) Molecular cloning of Bac7, a proline-and arginine-rich antimicrobial peptide from bovine neutrophils. *FEBS Lett.* **352,** 197–200.
44. Flodgard, H., Ostergaard, E., Bayne, S., Svendsen, A., Thomsen, J., Engels, M., and Wollmer, A. (1991) Covalent structure of two novel heterophil leucocyte-derived proteins of porcine and human origin. Neutrophil elastase homologues with strong monocyte and fiboblast chemotactic action. *Eur. J. Biochem.* **197,** 535–547.
45. Zanetti, M., Gennaro, T., and Romeo, D. (1995) Cathelicidins: a novel protein family with a common proregion and a variable C-terminal antimicrobial domain. *FEBS Lett.* **374,** 1–5.
46. vanAbel, R. J., Tang, Y. Q., Rao, V. S., Dobbs, C. H., Tran, D., Barany, G., and Selsted, M. E. (1995) Synthesis and characteracation of indolicidin, a tryptophan-rich antimicrobial peptide from bovine neutrophils. *J. Pept. Protein Res.* **45,** 401–409.
47. Seldsted, M. E., Tang, Y. Q., Morris, W. L., McGuire, P. A., Novotny, M. J., Smith, W., Henschen, A. H., and Collor, J. S. (1993) Purification, primary structures, and antibacterial activities of beta-defensins, a new family of antimicrobial peptides from bovine neutrophils. *J. Biol. Chem.* **268,** 6641–6648.
48. Pereira, H. A., Shafer, W. M., Pohl, J., Martin, L. E., and Spitznagel, J. K. (1990) CAP37, a human neutrophil-derived chemotactic factor with monocyte specific activity. *J. Clin. Invest.* **85,** 1468–1876.
49. Territo, M. C., Ganz, T., Selsted, M. E., and Lehrer, R. I. (1989) Monocyte chemotactic activity of defensins from human neutrophils *J. Clin. Invest.* **85,** 2017–2020.
50. Zhu, Q. Z., Singh, A. V., Esch, F., and Solomon, S. (1987) The coticostatic (anti-ACTH) and cytotoxic activity of peptides isolated from fetal, adult and tumor bearing lung. *J. Steroid Biochem.* **27,** 1017–1022.
51. Zasloff, M. (1987) Magainins, a class of antimicrobial peptides from *Xenopus* skin: characterization of two active forms and partial cDNA sequence of a precursor. *Proc. Natl. Acad. Sci. USA* **84,** 5449–5453.
52. Mignona, G., Simmaco, M., Kreil, G., and Barra, D. (1993) antibacterial and haemolytic peptides containing D-alloisoleucine from the skin of Bombina variegata. *CMBO J.* **12,** 4829–4832.

53. Lee, J. Y., Boman, A., Sun, C., Andersson, M., Jornvall, H., Mutt, V., and Boman, H. G. (1989) Antibacterial peptides from pig intestine: isolation of a mammalian cecropin. *Proc. Natl. Acad. Sci. USA* **86,** 9159–9162.
54. Hultmark, D., Steiner, H., Rasmuson, T., and Boman, H. G. (1980) Insect immunity: purification and properties of three inducible bactericidal proteins from hemolymph of immunized pupae of Hyalophora cecropia. *Eur. J. Biochem.* **106,** 7–16.
55. Hultmark, D. (1994) *Drosophila* as a model system for antibacterial peptides, *Antimicrobial Peptides* (Boman, H. G., ed.) *Ciba Found. Symp.* **186,** 107–119, Wiley, New York.
56. Lee, J. Y., Boman, A., Sun, C., Andersson, M., Jornvall, H., Mutt, V., and Boman, H. G. (1989) Antibacterial peptides from pig intestine: isolation of a mammalian cecropin. *Proc. Natl. Acad. Sci. USA* **86,** 9159–9162.
57. Eisenhauer, P. B., Harwig, S. S. S. L., and Leherer, R. I. (1992) Cryptdins: antimicrobial defensins of the murine small intestine. *Infect. Immun.* **60,** 3556–3565.
58. Jones, D. E. and Bevins, C. L. (1992) Paneth cells of the human small intestine express an antimicrobial peptide gene. *J. Biol. Chem.* **267,** 23,216–23,225.
59. Iwanaga, S., Muta, T., Shigenaga, T., Seki, M., Kawano, K., Katsu, T., and Dawabata, S. (1994) Structure-function relationships of tachyplesins and their analoques. *Ciba Found. Symp.* **186,** 160–174.
60. Maloy, W. L. and Kari, U. P. (1995) Structure-activity studies on magainins and other host defense peptides. *Biopolymers* **37,** 105–122.
61. Casteels-Josson, K., Capaci, T., Casteeels, P., and Tempst, P. (1993) Apidaecin multipeptide precursor structure: a putative mechanism for amplification of the insect antibacterial response. *EMBO J.* **12,** 1569–1578.
62. Casteels, P., Ampe, C., Jacobs, F., and Tempst, P. (1993) Functional and chemical characterization of hymenoptaecin, an antibaterial polypeptide that is infection-inducible in the honeybee *(Apis mellifera). J. Biol. Chem.* **268,** 7044–7054.
63. Kokryakov, V. N., Harwig, S. S., Panyutich, E. A., Shevchenko, A. A., Aleshina, G. M., Shamova, O. V., Korneva, H. A., and Lehrer, R. I. (1993) Protegrins: leukocyte antimicrobial peptides that combine features of corticostatic defensins and tachyplesins. *FEBS Lett.* **327,** 231–236.
64. Troxler, R. F., Offner, G. D., Xu, T., Banderspek, J. C., and Oppenheim, F. G. (1990) Structural relationship between human salivary histatins. *J. Dental Res.* **69,** 2–6.
65. Oppenheim, F. G., Xu, T., McMillian, F. M., Levitz, S. M., Diamond, R. D., Offner, G. D., and Troxler, R. F. (1988) Histatins, a novel family of histidine-rich proteins in human parotid secretion. solation, characterization, primary structure and fungistatic effects on Candida albicans *J. Biol. Chem.* **263,** 7472–7477.
66. Maloy, W. L. and Kari, U. P. (1995) Structure-activity studies on magainins and other host defense peptides. *Biopolymers* **37,** 105–122.
67. Kagan, B. L., Selsted, M. E., Ganz, T. and Lehrer, R. I. (1990) Antimicrobial defensin peptides form voltage-dependent ion-permeable channels in planer lipid bilayer membranes. *Proc. Natl. Acad. Sci. USA* **87,** 1570–1590.

68. Lehrer, R. I., Barton, A., Daher, K. A., Harwig, S. S. L., Ganz, T., and Selster, M. E. (1989) Interaction of human defensins with *Escherichia coli* mechanism of bactericidal activity. *J. Clin. Invest.* **84,** 553–561.
69. Rest, R. F., Cooney, M. H., and Spitznagel, J. K. (1977) Susceptibility of lipopolysaccharide mutants to the bactericidal action of juman neutrophil lysosomal fractions. *Infect. Immun.* **16,** 145–151.
70. Shafer, W. M., Casey, S. G., and Spitznagel, J. K. (1984) Lipid A and resistance of Salmonella typhimurium to antimicrobial granule proteins of human neutrophil granulocytes. *Infect. Immun.* **43,** 834–838.
71. Roland, K. L., Martin, L. E., Esther, C. R., and Spitznagel, J. K. (1993) Spontaneous pmrA mutants of Sallmonella typhimurium LT2 defines a new two-component regulatory system with a possible role in virulence. *J. Bact.* **175,** 4154–4164.
72. Farley, M. M., Shafer, W. M., and Spitznagel, J. K. (1988) Lipopolysaccharide structure determines ionic and hydrophobic binding of a cationic antimicrobial neutrophil granule protein. *Infect. Immun.* **56,** 1589–1592.
73. Groisman, E. A., Parra-Lopez, C., Salcedo, M., Lipps, C. J., and Heffron, F. (1992) Resistance to host antimicrobial peptides is necessary for Salmonella virulence. *Proc. Natl. Acad. Sci. USA* **89,** 11,939–11,943.
74. Roland, K. L., Esther, C. R., and Spitznagel, J. K. (1994) Isolation and characterization of a gene, pmrD, from *Salmonella typhimurium* LT2. *J. Bact.* **176,** 3589–3597.

I

Isolation and Characterization of Antibacterial Peptides

2

HPLC Methods for Purification of Antimicrobial Peptides

Michael E. Selsted

1. Introduction

The advent of high performance liquid chromatography (HPLC) has greatly accelerated the discovery, purification, and characterization of antimicrobial peptides. Virtually every modern study of an antimicrobial peptide includes or was preceded by a description of its purification. The increased pace of peptide discovery and characterization has resulted from the development of sophisticated column and solvent delivery technology over the last four decades. Interestingly, the modern methods described here derive directly from standard open-column ("low performance") chromatographic modalities. Therefore, it is not surprising that virtually every method used in traditional column chromatography has been adapted to high performance methods. These include gel filtration, ion-exchange, and reversed-phase chromatography methods described in this chapter.

The major advantages of HPLC over traditional low-pressure chromatographic methods derive from the fact that column matrices have been produced that enable the delivery of solvents through the stationary support at a high flow rate with relatively little resolution-defeating diffusion. Because HPLC supports are typically silica or polymeric in nature, they can be packed under high pressure into a column format in which the solvent volume is quite small compared to an open column of similar dimensions. As a result of the compressed nature of the matrix packing, the intrinsic resistance to solvent flow (back pressure) is substantially increased. Therefore, variable-speed hydraulic pumps are required for solvent delivery.

From: *Methods in Molecular Biology, Vol. 78: Antibacterial Peptide Protocols*
Edited by: W. M. Shafer, Humana Press Inc., Totowa, NJ

This chapter will concentrate on HPLC methods proven to be of particular value for the isolation of antimicrobial peptides. As a class of biomolecules, many of the known antimicrobial peptides are members of families composed of molecules that have high degrees of sequence identity ***(1,2)***. High levels of sequence identity have been demonstrated for mammalian myeloid ***(2,3)*** and enteric ***(4–7)*** defensins, ß-defensins ***(8–11)***, cecropins ***(12)***, magainins ***(13)***, insect defensins ***(14)***, and plant defensins ***(15)***, and the physical chemical characteristics of the peptides are predictably also quite similar. Therefore, the high resolving power of HPLC serves as a particularly important method for isolation and purification of antimicrobial peptides. In addition, since the purified molecule is routinely used in antimicrobial assays, it is critical that the peptide preparation being tested be devoid of artifactual (antimicrobial) components introduced during purification. In this regard, HPLC techniques can provide a valuable tool for generating highly pure preparations for characterizing the antimicrobial acitivities and mechanisms of antimicrobial peptides.

2. Materials

1. HPLC system:
   a. A programmable solvent delivery system capable of producing gradient mixtures of at least two solvents. Most research applications are adequately served by pumps capable of delivering 0.2–10 mL/min.
   b. A sample injector assembly with a sample loop of 1–2 mL capacity.
   c. Variable wavelength UV detector capable of monitoring from 190 to 300 nm.
   d. Peak-actuated fraction collector.
   e. Chart recorder or computerized data collection system.
2. Columns: HPLC columns, most commonly stainless steel jacketed, prepacked with resin supports formulated specifically for the separation methodology selected (*see* **Tables 1** and **2**).
   a. Size-exclusion columns.
   b. Cation-exchange columns.
   c. Reversed-phase columns.
3. Solvent degassing apparatus or sparging SpeedVac apparatus and helium source.
4. Centrifugal evaporator (e.g., SpeedVac, Savant Instruments, Holbrook, NY) with trap and vacuum pump.
5. Lyophilizer.
6. HPLC-grade water (purchased or prepare by glass distillation and scrubbed with organic sieve).
7. Buffers and ion pairing agents (*see* **Table 2**).

Table 1
Recommended Supports for HPLC of Antimicrobial Peptides

Chromatographic mode	Column	Peptides purified	Ref.
Gel permeation	TSK G3000PW	ß-defensins, indolicidin	**Fig. 2**
	Waters I-125	Somatostatin; ß-endorphin	***(30)***
	Poly LC Polyhydroxyethyl aspartamide	LHRH, insulin	
Cation-exchange	Bio-Sil TSK-CM-3-SW	Defensins	**Fig. 3**; ***(20)***
	Poly LC PolyCAT A (polyaspartate)	Defensins	
RP-HPLC	Wide pore (300 Å) C4, C8, C18	Defensins	**Fig. 5** ***(4,17,19,25,31,32)***
	(Various suppliers)	ß-defensins	**Fig. 4** ***(8-10,33)***
		Bactenecins	***(22,34)***
		Cecropins	***(23,35)***
		Magainins	***(21)***
		Indolicidin	***(16)***

Table 2
Examples of Solvents for HPLC

Mode	Column	Buffer A	Buffer B
Gel permeation	TSK G3000PW	0.1% aqueous TFA, 36% acetonitrile	NA
	Waters I-125	0.1% aqueous TFA, 40% acetonitrile	NA
Ion exchange	Bio-Sil TSK-CM-3-SW	50 m*M* phosphate buffer, 10% acetonitrile, pH 6.5	Buffer A + 3.5*M* NaCl
RP-HPLC	C4, C8, C18	0.1% aqueous TFA	0.1% TFA in acetonitrile
		0.13% aqueous HFBA	0.13% HFBA in acetonitrile
		10 m*M* TEAP, pH 4.0–6.0	40:60, Buffer A:acetonitrile
		0.1% ammonium acetate, pH 6.0	20:80, Buffer A:acetonitrile
		50 m*M* phosphoric acid	50:50, Buffer A:acetonitrile

3. Methods

3.1. HPLC Methods

3.1.1. Preparation of Mobile Phase Solutions

1. Removal of particulates: It is important to filter solvent solutions prepared from solid reagents (e.g., ammonium acetate). Unmodified HPLC-grade solvents do not require filtration prior to use. Furthermore, addition of liquid, HPLC-grade reagents such as ion-pairing acids (trifluroacetic acid, heptafluorobutyric acid, phosphoric acid) can be added to solvents and used without filtration.
   a. Use the highest available grade of solid reagent.
   b. Dissolve weighed solid reagent in HPLC-grade water (Solution or Buffer "A").
   c. If using a binary (or gradient) elution system, add the solid reagent to the solvent for Solution "B" in a manner which ensures solubility of the solid reagent. For example, Solvent A might be 0.1% (w/v) ammonium acetate in water. If the organic component for solution B is acetonitrile, make the final concentration of acetonitrile 50%. Add an equal volume of 0.2% (w/v) ammonium acetate in water. Solution B will then be 0.1% ammonium acetate in 50:50 water/acetonitrile.
2. Avoid "outgassing" by removing dissolved air from mobile phases. Dissolved gases can be problematic. Bubbles in lines can get trapped in pump check valves, dramatically altering pump performance. Air introduced into columns will emerge as a long series of "spikes" in the chromatogram, and will require extensive washing to eliminate gas from the column.
3. Vacuum degassing.
   a. Transfer solution to a vacuum-safe vessel (e.g., side-arm filtration flask).
   b. Degas solvents by application of a vacuum using an appropriately trapped water aspirator connected to a laboratory faucet. Most air is usually removed in 10 min for solvent volumes up to 4 L.
4. Helium sparging.
   a. Connect a clean piece of Teflon tubing to a helium tank equipped with a two-stage regulator.
   b. Cut the tubing 75 cm from the distal end and insert a disposable 0.45-μ filter unit (syringe type) in series with the tubing to eliminate particulates from the helium stream.
   c. Place the distal end of the sparging line into the solvent vessel and secure with tape so that the tubing end is at the bottom of the vessel.
   d. Initiate the flow of helium with brisk (but not violent) bubbling for 10 min. Dissolved gasses should now be adequately sparged from the solvent, and it is ready to use.
   e. *Note:* Degassing of solutions using either vacuum or sparging techniques should be repeated daily prior to using the HPLC system. Remember that both procedures, if used for extended periods, may alter the composition of the solvent by diminishing the content of highly volatile components. Such alternations in solvent composition will necessarily modify chromatographic elution profiles.

3.1.2. System Equilibration

1. Equilibrate the solvent delivery system to obtain a flat, stable UV baseline. Flow rate should be appropriate to the column dimension and the chromatographic mode being used: for a typical 0.5 × 25 cm (id) column, flow rates of 0.2–1.0 mL/min are typically used for gel filtration, intermediate flow rates (0.5–3.0 mL/min) for RP-HPLC, and higher flow rates are possible for ion-exchange-based separations.
2. Check for leaks at pump heads and at all unions.
3. Verify solvent delivery rates by measuring the rate of flow at 100% solvent A, 50% A/50% B%, and 100% B. This measurement is easily carried out using a 10-mL glass graduated cylinder and a stop watch (*see* **Note 1**).

3.1.3. Blank Runs

Record the detector output from two "blank runs," the first by running the solvent program without injection, and the second with injection of the solvent in which the peptide is dissolved. These two steps will allow an assesment of pump performance, the cleanliness of the column, and any "background" peaks that should be considered in analyzing subsequent chromatograms.

3.1.4. Sample Preparation

Regardless of the chromatographic mode, it is essential that the peptide-containing sample be free of particulates.

1. Clarify the sample by filtration (0.2-μ Teflon filter mounted on a syringe) or by centrifugation.
2. To avoid precipitation of samples on columns, test for the solubility of the sample over a range of solvent compositions representative of those to be generated in the gradient elution.

3.1.5. Sample Injection

1. Using the sample syringe, wash the sample loop (in the "load" position) with 5 vol of the initial buffer, the final buffer, and then the initial buffer.
2. To avoid loss of valuable starting material, fill the loop with a volume of sample not exceeding 50% of the loop capacity.
3. Quickly flip the injector valve to the inject position and leave in that position until the end of the run.
4. Simultaneously initiate the gradient program (if applicable), start fraction collection, and begin recording the UV elution profile.

3.1.6. Fraction Collection

Eluent collection may be time-based, peak-actuated, or a combination. Many fraction collectors are designed for electronic interface with UV detectors for peak-based collection.

1. If peak-based collection is used, remember to enter a collection delay (programmed at the fraction collector) to correct for the volume between the UV flow cell and the tubing outlet at the fraction collector.
2. Tubes used for fraction collection should be appropriate to the scale and mode of separation.
3. For quantities < 5–10 µg, it is preferable to collect effluent in siliconized glass or polypropylene tubes.

3.2. Mode-Specific Protocols for Purification of Antimicrobial Peptides

It should be recognized that peptides and proteins all have unique personalities. The efficient purification of antimicrobial peptides will therefore benefit from a general understanding of the common biochemical characteristics of this group of molecules. In this regard, virtually all known antimicrobial peptides are small proteins (<10 kDa), a feature that may facilitate purification by size exclusion techniques; they are generally cationic and are thus separable from anionic species that comprise the majority of all proteins; and finally, they are nearly all readily separable by reversed-phase techniques. Specific protocols employing each of these modalities are presented in the following sections. The covalent structures of prototype molecules used in the examples are shown in **Fig. 1**.

3.2.1. Size Exclusion Chromatography

This mode of separation (also called gel filtration or gel permeation) fractionates mixtures as a function of mean molecular radius and is particularly valuable as a first step in fractionating peptides in crude extracts. The main limitation of the method is the relatively small capacity of HPLC-size exclusion columns (1–3% of column volume). An example of the method is the partial purification of ß-defensins and indolicidin from bovine neutrophils.

3.2.1.1. Sample

Lyophilized, 10% acetic acid extract of cytoplasmic granules from 6×10^8 peripheral blood bovine neutrophils *(8)*.

1. Gently dissolve protein in 1 mL of the eluant buffer: 0.1% TFA, 36% acetonitrile (**Tables 1** and **2**).
2. After maximum dissolution, centrifuge the sample at 20,000*g* at room temperature, and transfer the supernatant to a clean tube.

3.2.1.2. Chromatography

1. Connect two 7.5 × 300-mm and a 7.5 × 75-mm guard columns packed with TSK G3000PW (**Table 1**) in series.

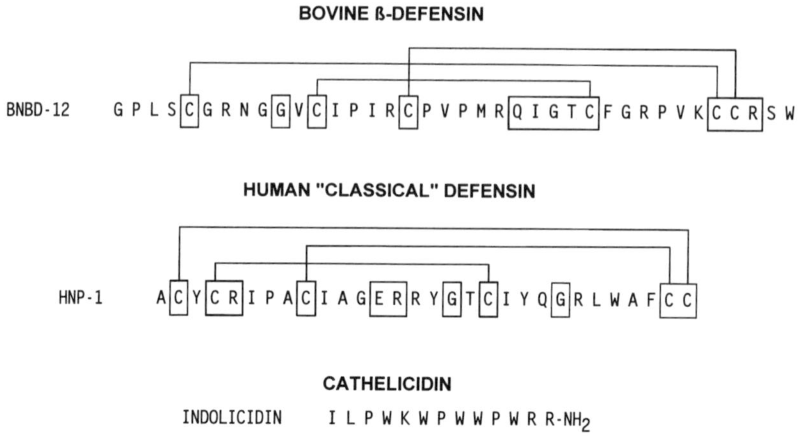

Fig. 1. Covalent structures of neutrophil antimicrobial peptides referenced in HPLC protocols. Conserved residues in ß-defensins and defensins are enclosed in boxes, and the disulfide motifs are indicated by lines connecting cysteines.

2. Using one pump only, equilibrate the column in 0.1% TFA/36% acetonitrile at 0.5 mL/min for 30–60 min or until a very stable baseline is obtained while monitoring the eluant at 280 nm (0.1 AUFS) and/or at 220 nm (1.0 AUFS).
3. Inject 100 µL of sample and collect using either time-based or peak-actuated collection. As shown in **Fig. 2**, early peaks are not separated to baseline, but later peaks give nearly baseline separation. ß-defensins ***(8)*** are in pool D, and indolicidin ***(16)*** is in pool E (*see* **Note 2**).

3.2.2. Cation-Exchange HPLC

Nearly all known antimicrobial peptides are cationic, making cation exchange chromatography an attractive mode of separation. One advantage of the method, compared to size-exclusion HPLC, is the large capacity of most resins, and the relatively high resolution that can be achieved. The relative disadvantage of this technique, compared to RP-HPLC, is the somewhat lower resolving capacity of this technique, and the requirement for carrying out an additional desalting step prior to testing of samples that are eluted with non-volatile salt solutions.

3.2.2.1. Sample

Lyophilized, low molecular weight fraction (<10,000 Dalton) obtained by BioGel P-10 column (Bio-Rad, CA) chromatography of a 10% acetic acid extract of 1×10^{10} rabbit peritoneal neutrophils ***(19,20)***.

1. Dissolve lyophilate in 20 mL of filtered 50 m*M* sodium phosphate, pH 6.7, containing 10% acetonitrile (Buffer A).

2. After maximum dissolution, centrifuge the sample at 20,000*g* at room temperature, and transfer the supernatant to a clean tube.

3.2.2.2. Chromatography

1. Attach a 21.5 × 150-mm Bio-Sil TSK-CM-3-SW column (Bio-Rad, Hercules, CA) (weak cationic exchanger; **Table 1**) to the system.
2. Equilibrate the column with 10–15 column volumes of Buffer A at 6 mL/min.
3. Apply a 0–100% gradient of buffer B (3.5*M* NaCl in buffer A) at 6 mL/min. The gradient should be developed in 10–15 min.
4. Wash the column with 10 column volumes of buffer B at 6 mL/min.
5. Re-equilibrate the column in buffer A until a flat baseline is obtained at A_{220} (1.0 AUFS).
6. Inject 10 mL of the sample 1.0 mL at a time, with a 1-min interval between injections.
7. Wait for at least 5 min after the last injection peak emerges or until a stable A_{220} baseline (1.0 AUFS) is obtained. Then apply a 0–100% gradient of buffer B in 100 min. (**Fig. 3**).
8. Collect 2 min (12-mL) fractions. Rabbit neutrophil defensins NP-3A, 3B, 4, and 5 are resolved to baseline, and NP-1 and 2 elute together (**Fig. 2**). A subsequent RP-HPLC step of the cation-exhange fractionated material readily resolves NP-1 and NP-2 (*see below*).

3.2.3. RP-HPLC

This is among the most powerful bioseparation methods, and is a mainstay among techniques for purification of antimicrobial peptides ***(8,9,21–25)***. The most common column packings are silica or polymeric supports to which straight chain hydrocarbons ranging from C4–C18 are bonded. Separation is predominantly driven by hydrophobic interactions of the solute with the packing. Elution of the adsorbed peptide is typically accomplished by gradient elution using water-miscible organic solvents. The "selectivity" of RP-HPLC separations can be substantially modified by the use of different ion-pairing reagents ***(26)***. Among those commonly used are: trifluoroacetic acid (TFA), heptafluorobutyric acid (HFBA), ammonium acetate, triethylammonium phosphate (TEAP), and phosphoric acid. The RP-HPLC example given is for the purification of bovine neutrophil ß-defensins ***(8)***.

3.2.3.1. Sample

Lyophilized, pooled fractions cooresponding to pool D in **Fig. 2**.

1. Gently dissolve sample in 1 mL of filtered 5% acetic acid.
2. After maximum dissolution, centrifuge the sample at 20,000*g* at room temperature, and transfer the supernatant to a clean tube.

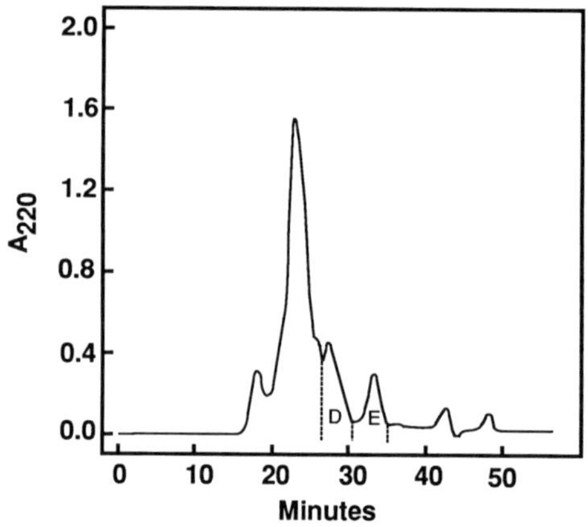

Fig. 2. Size exclusion HPLC of bovine neutrophil granule extract separated on a TSK G3000PW column. Pools D and E contain ß-defensins and indolicidin, respectively.

3.2.3.2. Chromatography

1. Equilibrate a 1 × 25 cm Vydac (The Separations Group, Hesperia, CA) C18 column (300-Å pore size) in aqueous 0.1% TFA (solvent A) at 3 mL/min until a flat baseline is obtained at A_{230} (0.5 AUFS).
2. Remember to perform a blank run (*see* **Subheading 3.1.3.**), and re-equilibrate the column.
3. Inject entire 1-mL sample, and initiate a linear gradient of 0.1% TFA in acetonitrile (Buffer B) at 2%/min for 10 min (to 20% acetonitrile). Then apply a shallow acetonitrile gradient: 20–45% acetonitrile at 0.33%/min.
4. Collect samples with monitoring at 230 nm. Thirteen ß-defensins elute in the positions indicated in **Fig. 4**.
5. Take each ß-defensin-containing peak (**Fig. 4**) to dryness in a SpeedVac centrifugal evaporator.
6. Dissolve each dried sample in 2 mL of 0.13% heptafluorobutyric acid (HBFA).
7. Equilibrate the C18 column in water containing 0.13% HFBA.
8. Perform blank run from 20–70% B (B = 0.13% HFBA in acetonitrile) over 50 min, and re-equilibrate column at 20% B.
9. Inject 1 mL (50%) of dissolved sample (from **step 6**) and apply a 20–70% B linear gradient at 1%/min. Collect peaks as above.
10. Lyophilize samples and analyze for purity on acid-urea PAGE ***(27)*** and analytical RP-HPLC (*see* **Note 3**).

3.2.4. Combining Cation-Exchange and RP-HPLC

The combination of these two chromatographic modes provides a powerful approach to the purification of antimicrobial peptides. The purification of rab-

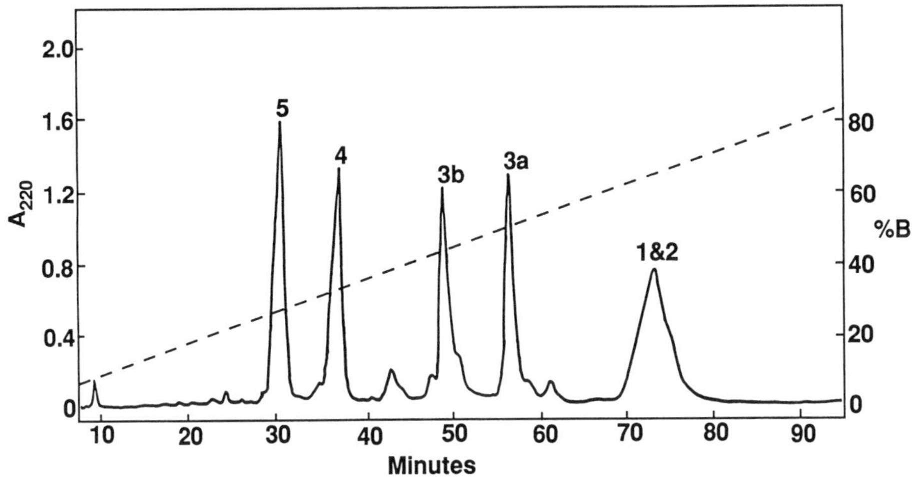

Fig. 3. Cation exchange chromatogram of rabbit neutrophil defensins 1–5 *(20)* resolved on a Bio-Sil TSK CM-3-SW column. Percent of B buffer in elution gradient is indicated with dashed line.

bit neutrophil defensins is greatly facilitated by the sequential application of cation exchange (**Fig. 3**) and RP-HPLC *(20)*. By directly subjecting an ion-exchange purified sample, in this case rabbit defensins NP1–5, to a RP-HPLC step, the peptide can be simultaneously desalted (since the phosphate and sodium chloride are not retained during loading) and further purified.

3.2.4.1. Samples

Fractions corresponding to each of the five labeled peaks in **Fig. 3**.

1. Subject samples to 15 min of vacuum concentration in a Speed Vac evaporator to remove most of the acetonitrile contained in the cation-exchange chromatography solvent.
2. Acidify each sample by the addition of acetic acid to a final concentration of 5%.

3.2.4.2. Chromatography

1. Equilibrate a 1 × 25 cm Vydac C18 column in aqueous 0.1% TFA (Buffer A) at 3 mL/min until a stable baseline (A_{220}, 1.0 AUFS) is obtained. Buffer B is 0.1% TFA in acetonitrile.
2. Perform blank run (0–45% B at 1%/min) and re-equilibrate column.
3. Stop the pumps.
4. With gloved hands, carefully place the Buffer A inlet line in a vessel containing up to 200 mL of solution corresponding to one of the peaks in **Fig. 3**. Do not allow any air bubbles into the system.

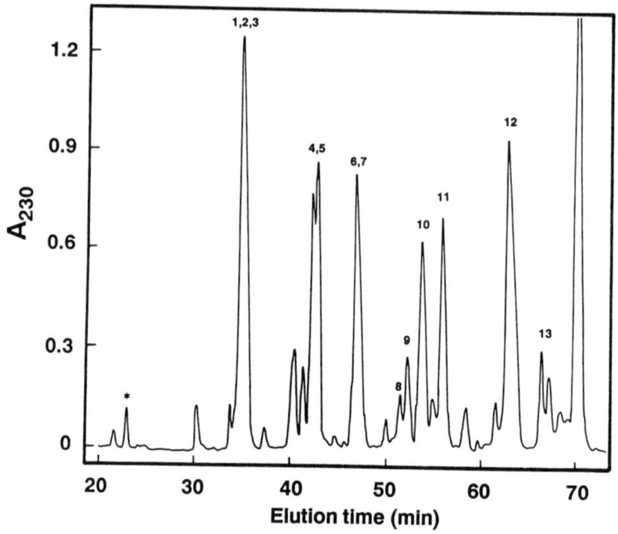

Fig. 4. C18 RP-HPLC chromatogram of bovine neutrophil ß-defensins. Material loaded on this column was Pool D from chromatogram shown in **Fig. 2** (reproduced with permission from **ref.** ***8***).

5. Start pump A, and pump the sample from the vessel directly onto the column at 2 mL/min.
6. Carefully monitor the level of the sample solution in the vessel. As the level approaches the opening of the inlet tubing, begin washing the the vessel with 1–5-mL aliquots of Buffer A. Continue washing until the A_{220} tracing returns to a stable baseline.
7. Stop the pump.
8. Return the inlet tubing to the Buffer A reservoir, and restart the pumps.
9. Equilibrate the flow rate to 3 mL/min.
10. Apply a 0–45% gradient of Buffer B at 1%/min.
11. Monitor the column eluant at 230 nm, and collect the fractions corresponding to each major peak.

3.2.5. Determination of Peptide Purity

In most instances, the apparent purity of a peptide (as evidenced by RP-HPLC elution of a single symmetrical peak) can be confirmed by analytical RP-HPLC combined with acid-urea PAGE ***(29)***. Analytical RP-HPLC should be carried out on a narrow (2–5 mm) C18 column using a 0–60% acetonitrile gradient. The ion-pairing reagent chosen should be different than that used in the final purification step. Analysis on a 12.5% acid-urea

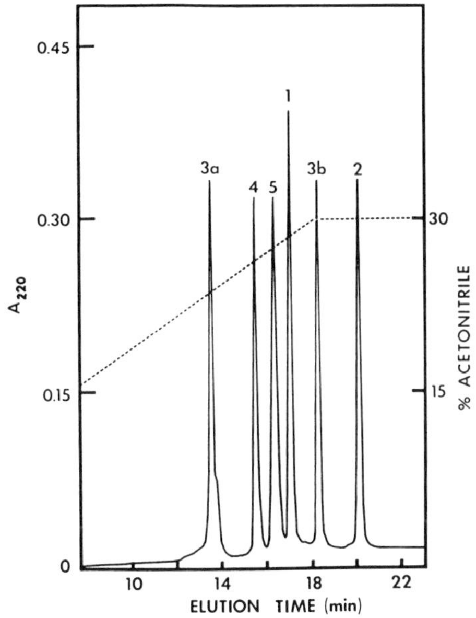

Fig. 5. Analytical RP-HPLC of purified rabbit neutrophil defensins. Acetonitrile gradient is indicated by dashed line (reproduced with permission from **ref. *19***).

polyacrylamide gel *(27)* is often helpful in further confirming the homogeneity of the preparation.

3.2.5.1. Sample

The peptide should be dissolved in the solvent that it will be stored in, or alternatively in a volatile buffer (e.g., 1% acetic acid) that can be removed by lyophilization.

1. Set the UV monitor to a wavelength between 207 and 230 nm. The sensitivity increases as the wavelength is lowered.
2. Perform a blank run using the buffer in which the peptide is dissolved.
3. Inject an amount of peptide that gives a 50–90% full scale recorder deflection.
4. Collect UV absorbing material using a peak-actuated fraction collector. Evidence of peptide purity is obtained if the chromatogram shows a a single symmetrical peak, and there is no evidence of earlier or later eluting material. A mixture of RP-HPLC-purified neutrophil defensins, injected on an analytical C18 column is shown in **Fig. 5**.
5. To verify purity, subject purified peptides to a second RP-HPLC step using a different ion pairing agent (e.g., HFBA, TEAP, or ammonium acetate), and/or evaluate homogeneity by AU-PAGE, or mass spectroscopy.

4. Notes

1. Make sure the system is clean! With the column (and precolumn) removed from the system, you should treat stainless steel systems with 50% nitric acid by pumping it directly through the pumps, injector loop, lines, and UV detector to waste. Usually, 50–100 mL is adequate to remove deposits from the system and to repassivate stainless steel. Wash the system thoroughly with HPLC-quality water, check for leaks, equilibrate the system with desired solvents, and reattach the column.
2. The maximum single injection volume may be increased if separation is adequate (e.g., *see* resolution of indolicidin, pool E in **Fig. 2**). Traditional open column gel filtration ***(17,18)*** is often necessary for larger scale purification using gel permeation methods.
3. The use of more than one ion-pairing agent provides "multidimensional" RP-HPLC separations. The relative retention times and order of peptide elution can be altered by simply changing the solvent modifier. Alternating between TFA and HFBA, or between TFA and 0.1% ammonium acetate ***(28),*** can markedly facilitate difficult separations. It should be noted that, although TFA, HFBA, and ammonium acetate are volatile salts, they have very different behaviors during lyophilization. For purification of antimicrobial peptides, it is recommended that the final purification step utilize TFA as the ion pair, as this acid is easily removed by lyophilization. Other volatile modifiers are not so easily evaporated, and their presence at unknown levels may influence the antimicrobial properties of the purified peptide.

Acknowledgments

I would like to acknowledge the support of NIH grants AI-22931 and AI31696, the UC Tobacco-Related Disease Research Grant RT83, and Biosource Technologies.

References

1. Boman, H. G. (1995) Peptide antibiotics and their role in innate immunity. *Annu. Rev. Immunol.* **13,** 62–92.
2. Martin, E., Ganz, T., and Lehrer, R. I. (1995) Defensins and other endogenous peptide antibiotics of vertebrates. *J. Leukocyte Biol.* **58,** 128–136.
3. Lehrer, R. I., Lichtenstein, A. K., and Ganz, T. (1993) Defensins: antimicrobial and cytotoxic peptides of mammalian cells. *Annu. Rev. Immunol.* **11,** 105–128.
4. Ouellette, A. J., Hsieh, M. M., Nosek, M. T., Cano-Gauci, D. F., Huttner, K. M., Buick, R. N., and Selsted, M. E. (1994) Mouse paneth cell defensins: primary structures and antibacterial activities of numerous cryptdin isoforms. *Infect. Immun.* **62,** 5040–5047.
5. Selsted, M. E. and Ouellette, A. J. (1995) Defensins in granules of phagocytic and non-phagocytic cells. *Trends in Cell Biology* **5,** 114–119.

6. Jones, D. E. and Bevins, C. L. (1992) Paneth cells of the human small intestine express an antimicrobial peptide gene. *J. Biol. Chem.* **267,** 23,216–23,225.
7. Jones, D. E. and Bevins, C. L. (1993) Defensin-6 mRNA in human Paneth cells: implications for antimicrobial peptides in host defense of the human bowel. *FEBS Lett.* **315,** 187–192.
8. Selsted, M. E., Tang, Y.-Q., Morris, W. L., McGuire, P. A., Novotny, M. J., Smith, W., Henschen, A. H., and Cullor, J. S. (1993) Purification, primary structures, and antibacterial activities of ß-defensins, a new family of antimicrobial peptides from bovine neutrophils. *J. Biol. Chem.* **268,** 6641–6648.
9. Diamond, G., Zasloff, M., Eck, H., Brasseur, M., Maloy, W. L., and Bevins, C. L. (1991) Tracheal antimicrobial peptide, a cysteine-rich peptide from mammalian tracheal mucosa: peptide isolation and cloning of a cDNA. *Proc. Natl. Acad. Sci. USA* **88,** 3952–3956.
10. Schonwetter, B. S., Stolzenberg, E. D., and Zasloff, M. A. (1995) Epithelial antibiotics induced at sites of inflammation. *Science* **267,** 1645–1648.
11. Bensch, K. W., Raida, M., Magert, H.-J., Schulz-Knappe, P., and Forssmann, W.-G. (1995) hBD-1: a novel ß-defensin from human plasma. *FEBS Lett.* **368,** 331–335.
12. Boman, H. G., Faye, I., Gudmundsson, G. H., Lee, J.-Y., and Lidholm, D.-A. (1991) Cell-free immunity in *Cecropia*. A model system for antibacterial proteins. *Eur. J. Biochem.* **201,** 23–31.
13. Bevins, C. L. and Zasloff, M. (1990) Peptides from frog skin. *Ann. Rev. Biochem.* **59,** 395–414.
14. Hoffmann, J. A. and Hetru, C. (1992) Insect defensins: inducible antibacterial peptides. *Immunol. Today* **13,** 411–415.
15. Broekaert, W. F., Terras, F. R. G., Cammue, B. P. A., and Osborn, R. W. (1995) Plant defensins: novel antimicrobial peptides as components of the host defense system. *Plant Physiol.* **108,** 1353–1358.
16. Selsted, M. E., Novotny, M. J., Morris, W. L., Tang, Y.-Q., Smith, W., and Cullor, J. S. (1992) Indolicidin, a novel bactericidal tridecapeptide amide from neutrophils. *J. Biol. Chem.* **267,** 4292–4295.
17. Selsted, M. E., Miller, S. I., Henschen, A. H., and Ouellette, A. J. (1992) Enteric defensins: antibiotic peptide components of intestinal host defense. *J. Cell Biol.* **118,** 929–936.
18. Harwig, S. S. L., Ganz, T., and Lehrer, R. I. (1994) Neutrophil defensins: purification, characterization, and antimicrobial testing. *Meth. in Enzymol.* **236,** 160–172.
19. Selsted, M. E., Szklarek, D., and Lehrer, R. I. (1984) Purification and antibacterial activity of antimicrobial peptides of rabbit granulocytes. *Infect. Immun.* **45,** 150–154.
20. Lichtenstein, A., Ganz, T., Selsted, M. E., and Lehrer, R. I. (1986) *In vitro* tumor cell cytolysis mediated by peptide defensins of human and rabbit granulocytes. *Blood* **68,** 1407–1410.

21. Zasloff, M. (1987) Magainins, a class of antimicrobial peptides from *Xenopus* skin: isolation, characterization of two active forms, and partial cDNA sequence of a precursor. *Proc. Nat. Acad. Sci. USA* **84,** 5449–5453.
22. Gennaro, R., Skerlavaj, B., and Romeo, D. (1989) Purification, composition, and activity of two bactenecins, antibacterial peptides of bovine neutrophils. *Infect. Immun.* **57,** 3142–3146.
23. Lee, J.-Y., Boman, A., Chuanxin, S., Andersson, M., Jornvall, H., Mutt, V., and Boman, H. G. (1989) Antibacterial peptides from pig intestine: Isolation of a mammalian cecropin. *Proc. Nat. Acad. Sci. USA* **86,** 9159–9162.
24. Agerberth, B., Lee, J.-Y., Bergman, T., Carlquist, M., Boman, H. G., Mutt, V., and Jornvall, H. (1991) Amino acid sequence of PR-39. Isolation from pig intestine of a new member of the family of proline-arginine-rich antibacterial peptides. *Eur. J. Biochem.* **202,** 849–854.
25. Ganz, T., Selsted, M. E., Szklarek, D., Harwig, S. S. L., Daher, K., Bainton, D. F., and Lehrer, R. I. (1985) Defensins. Natural peptide antibiotics of human neutrophils. *J. Clin. Invest.* **76,** 1427–1435.
26. Guo, D., Mant, C. T., and Hodges, R. S. (1987) Effects of ion-pairing reagents on the prediction of peptide retention in reversed-phase high performance liquid chromatography. *J. Chrom.* **386,** 205–222.
27. Selsted, M. E. and Becker, H. W., III. (1986) Eosin Y: a reversible stain for detecting electrophoretically resolved protein. *Anal. Biochem.* **155,** 270–274.
28. Anderson, J. K. and Mole, J. E. (1983) Adaptation of reverse-phase high-performance liquid chromatography for the isolation and sequence analysis of peptides from plasma amyloid p-component, in *High-Performance Liquid Chromatography of Proteins and Peptides* (Hearn, M. T. W., Regnier, F. E., and Wehr, C. T., eds.), Academic, New York, pp. 29–37.
29. Selsted, M. E. (1993) Investigational approaches for studying the structures and biological functions of myeloid antimicrobial peptides, in *Genetic Engineering: Principles and Methods* (Setlow, J. K., ed.), Plenum, New York, pp. 131–147.
30. Bennett, H. P. J., Browne, C. A., and Solomon, S. (1983) α-N-Acetyl-ß-Endorphin1-26 from the neurointermediary lobe of the rat pituitary: isolation, purification, and characterization by high-performance liquid chromatography, in *High-Performance Liquid Chromatography of Proteins and Peptides* (Hearn, M. T. W., Regnier, F. E., and Wehr, C. T., eds.), Academic, New York, p. 253–261.
31. Eisenhauer, P. B., Harwig, S. S. L., Szklarek, D., Ganz, T., Selsted, M. E., and Lehrer, R. I. (1989) Purification and antimicrobial properties of three defensins from rat neutrophils. *Infect. Immun.* **57,** 2021–2027.
32. Selsted, M. E., Brown, D. M., DeLange, R. J., and Lehrer, R. I. (1983) Primary structures of MCP-1 and MCP-2, natural peptide antibiotics of rabbit lung macrophages. *J. Biol. Chem.* **258,** 14,485–14,489.
33. Harwig, S. S. L., Swiderek, K. M., Kokryakov, V. N., Tan, L., Lee, T. D., Panyutich, E. A., Aleshina, G. M., Shamova, O. V., and Lehrer, R. I. (1994) Gallinacins: cysteine-rich antimicrobial peptides of chicken leukocytes. *FEBS Lett.* **342,** 281–285.

34. Romeo, D., Skerlavaj, B., Bolognesi, M., and Gennaro, R. (1988) Structure and bactericidal activity of an antibiotic dodecapeptide purified from bovine neutrophils. *J. Biol. Chem.* **263,** 9573–9575.
35. Hultmark, D., Steiner, H., Rasmuson, T., and Boman, H. G. (1980) Insect immunity. Purification and properties of three inducible bactericidal proteins from hemolymph of immunized pupae of *Hyalophora cecropia*. *Eur. J. Biochem.* **106,** 7–16.

3

Strategies for the Isolation and Characterization of Antimicrobial Peptides of Invertebrates

Charles Hetru and Philippe Bulet

1. Introduction

Resistance of bacteria to antibiotics has become one of the main problems in human health. In addition, in agronomy, microbial diseases are largely responsible for the decrease in agricultural production. The discovery of new antibiotic families is a way to circumvent such problems and antimicrobial peptides may represent a new type of such antibiotics.

In animals, antimicrobial peptides are important effectors of the innate immune response (nonadaptive immunity). In mammals, they are produced by neutrophils or macrophages and kill microbial invaders in a first barrier of host defense. The acquired and specific immune responses with production of antibodies occur later. In insects, there is no specific and adaptive immunity but only an innate response that includes cellular and humoral factors. The cellular response consists mainly of phagocytosis and encapsulation. The humoral immune response includes the rapid synthesis of a battery of antimicrobial peptides.

A recent review lists the antimicrobial peptides/polypeptides isolated from insects and shows the rapid increase in the number of molecules that have been characterized *(1)*. To date, from only 22 species, more than 100 different peptides/polypeptides have been fully characterized. Invertebrates, which include the insects, present an extreme diversity and a potential source of a large variety of antimicrobial substances. Since the discovery of the first antibacterial peptide in insects in 1981 *(2)*, a large variety of techniques have been used to isolate, purify, and characterize such molecules.

From: *Methods in Molecular Biology, Vol. 78: Antibacterial Peptide Protocols*
Edited by: W. M. Shafer, Humana Press Inc., Totowa, NJ

In this chapter, we describe a number of methods for the identification, purification, and characterization of antimicrobial peptides/polypeptides. Several of the topics of the present chapter have certainly been covered in earlier reviews; however, to be complete and for the reader's autonomy, we have chosen to describe all the techniques needed. We have focused our descriptions and advice to specific aspects of invertebrate antimicrobial peptides/polypeptides: induction to the antimicrobial peptides, preparation of the samples, and, as the material will be obtained in very small amounts, on modified methods for chemical characterizations and antimicrobial tests.

The principal group of invertebrates that has been used until now, for the isolation of antimicrobial molecules, is the insect class ***(3)***. In this class, most of the antimicrobial peptides are not present in the hemolymph of normal animals, but are induced by an injury or an injection of microbes ***(4)***. Thus the first operation, prior to extraction and purification, is the induction of the production of antimicrobial peptides by the animals.

Various kind of inducers have been injected: living bacteria from 10^4–10^6 cells per animal ***(5–11)***, heat-killed microbes ***(12)***, and components of the cell-wall of bacteria ***(13,14)***.

However, to obtain a more complete induction of antimicrobial peptides, the injection of a mixture of living Gram-negative and Gram-positive bacteria at sublethal doses is recommended.

The main source of antimicrobial peptides in invertebrates is hemolymph, and the first step after collection is centrifugation to separate plasma from hemocytes. In order to remove from plasma, compounds other than the peptides of interest, several treatments have been proposed: heat treatment ***(6)*** or plasma acidification ***(15)***.

For very small animals, to simplify the collection of material the protein extraction is directly performed from the total body of the animals.

Recently, solid-phase extraction (SPE) on reversed-phase (C18) has been used to prepare peptide samples ***(11,16–21)***. This procedure consists of a kind of crude chromatography. Before loading on this support, the sample should be acidified, typically with trifluoroacetic acid (TFA) or acetic acid. A sequential elution with low, medium, and high percentage of acetonitrile (or methanol) in acidified water leads to a prepurification of the sample. Salts, sugars, and most hydrophilic proteins are eliminated during the washing cycle, whereas lipids and most hydrophobic proteins are retained on the solid-phase.

Antimicrobial peptides from arthropods have also been isolated from hemocytes ***(22,23)*** that are homogenized, centrifuged, and the supernatant is extracted as for the plasmatic fraction.

In the case of cultured cells, the medium can be submitted directly to purification without any concentration ***(24,25)***.

After sample preparation, the extracts are concentrated by lyophilization or in a vaccum centrifuge and then resuspended in the appropriate buffers for purification.

The principle mode of purification used for antimicrobial peptides is chromatography mainly on reverved-phase and size-exclusion columns.

The characterization is mainly performed by combination of sequencing by automated Edman degradation, mass spectrometry analysis, and enzymatic cleavage.

Monitoring of antibacterial activity by the paper disk method is the oldest protocol available and is still currently used for antibiograms (medical and pharmaceutical diagnostics). A variation of this method, often used, is the inhibition zone assay ***(5,8,9,12,26–31)***. Recently, a sensitive liquid growth inhibition assay has been described ***(17,20)***.

Antifungal assay against filamentous fungi can now be easily conducted with a very good sensitivity also in a liquid growth inhibition assay.

2. Materials

2.1. Insect Immunization

1. Gram-positive bacteria: *Micrococcus luteus*.
2. Gram-negative bacteria: *Escherichia coli* 1106 or any nonpathogen wild type.
3. Luria Broth medium (LB): 15.5g of Millers's modification of Luria Broth, Gibco-BRL; 1L of H_2O, pH 7.4 or 1% bactotrypton, 0.5% yeast extract, 0.9% NaCl w/v.
4. Eppendorf tubes, 1.5- or 2.5-mL for bacterial dilutions or to put the bacterial pellet into the cap.
5. Hamilton syringe (5 or 10 µL) or stainless steel needle (ultrafine).

2.2. Extraction

1. Phenylthiourea (Sigma, St. Louis, MO; PTU: stock solution 20 m*M* in ethanol).
2. Aprotinin (protease inhibitor; Sigma).
3 Trifluoroacetic acid, sequenal grade (Pierce, Rockford, IL).
4. Filter units, 0.8-µm membrane (Millex unit, Millipore, Bedford, MA).

2.3. Purification of Antimicrobial Peptides

1. Solid-phase extraction cartridges (Sep-Pak C18 cartridges, Waters, Milford, MA). Several sizes are available according to the quantity of extract.
2. Methanol for HPLC (Carlo Erba, Rodano, Italy).
3. Acetonitrile for HPLC (Merck, Rahway, NJ).
4. HPLC water or any ultrapure water (MilliQ water, Millipore).
5. Polypropylene tubes Minisorp (NUNC Immuno tubes, 75 × 12 mm, Roskilde, Denmark).
6. HPLC columns: Porosity 300 Å, granulometry 7 µm, reversed-phase C8 or C18, analytical columns (2.1–4.6 mm id).

7. Size-exclusion column (SEC 2000 and SEC 3000, Beckman, Fullerton, CA) and a precolumn (Beckman).
8. HPLC system: pump (one or two), UV detector (detection at 225 nm) with two output (analogical for paper recorder and digital for computer) and a paper recorder.
9. Centrifuge vacuum drier (Speed Vac, Savant, Hicksville, NY).

2.4. Microsequencing Analysis

1. Automated Edman degradation of the pure peptide and detection of phenylthiohydantoin derivatives are performed on a pulse liquid automatic sequenator (e.g., Perkin Elmer Applied Biosystems, model 473A). Reagent and solvants are purchased to manufacturer (Perkin Elmer, Applied Biosystem Division, Norwalk, CT).

2.5. Mass Spectrometry

1. Electrospray ionization mass spectrometer with an electrostatic ion spray source operating at atmospheric pressure followed by a quadrupole mass analyzer (mass range 1–4000, scanning from *m/z* 500 to *m/z* 1500 in 10 s) (VG Biotech BioQ mass spectrometer, Manchester, UK).
2 Multichannel analyzer as data system.
3. Calibration with heart myoglobin.
4. Acetic acid and methonol for analysis quality.
5. Matrix-assisted laser desorption/ionisation time-of-flight mass spectrometer (Bruker, Bremen, Germany).
6. α-cyano-4-hydroxycinnamic acid (Sigma).
7. Moderate vacuum pump (membrane pump).
8. Standard peptides for calibration: angiotensin II, ACTH 18-39 and bovine insulin.

2.6. Enzymatic Cleavage

1. Water bath (37°C).
2. Polypropylene tubes.
3. Protease of sequencing grade (Proteases from Boehringer, Mannheim or residue-specific protease kit from Takara, Japan).
4. Reaction buffers:
   a. Lysyl endoproteinase (*Achromobacter* protease I), 10–25 m*M* Tris-HCl, pH 8.0, 0.01% Tween-20 with or without 1 m*M* EDTA.
   b. Arginyl endoproteinase, 10 m*M* Tris-HCl, pH 8.0, 0.01% Tween-20.
   c. Trypsin, 10 m*M* Tris-HCl, pH 8.0, 0.01% Tween-20, 10 m*M* $CaCl_2$.
   d. *Staphylococcus aureus* V8 proteinase, 50 m*M* ammonium carbonate, pH 7.8, 0.01% Tween-20.
   e. Asparaginyl endopeptidase, 20 m*M* sodium acetate buffer, pH 5.0, 0.01%, Tween-20, 1 m*M* DTT, 1 m*M* EDTA.

f. Pyroglutamate aminopeptidase, 100 m*M* sodium phosphate buffer, pH 8.0, 10 m*M* EDTA, 5% glycerol, 5 m*M* dithiothreitol.
g. Carboxypeptidase Y and P, 50 m*M* sodium citrate buffer, pH 6.0 (for Y) and pH 4.0 (for P).

2.7. Bioassays

2.7.1. Preparation of Spore Suspension of Neurospora crassa

1. Six cereal agar: 20 g of six cereal instant flakes (Nestlé) and 15 g agar in 1 L of sterilized water.
2. Autoclaved water.
3. Autoclaved 50% glycerol.
4. Sterile microtubes (Eppendorf tubes).
5. Broad spatula.
6. Glass funnel plugged with glass wool wrapped in aluminium foil and autoclaved.

2.7.2. Antifungal and Antibacterial Tests

1. Potato dextrose broth (DIFCO): 12 g PDB for 1 L water.
2. Tetracycline (Sigma) stock solution: 10 mg/mL in DMSO.
3. Cefotaxim (Sigma) stock solution: 100 µg/mL in water.
4. Luria Bertani's rich medium: 1% bactotrypton, 0.5% yeast extract, 0.9% NaCl w/v.
5. Poor broth medium: 1% bactotrypton, 0.9% NaCl w/v.
6. 96-well microtiter plates for cell culture (Nunc, Wiesbaden-Biebrich, Germany).
7. Microplate reader.

2.8. Reduction and Alkylation

1. Reaction buffer: 0.5*M* Tris-HCl, 2 m*M* EDTA, pH 7.5 containing 6*M* guanidinium chloride.
2. Dithiothreitol stock solution: 2.2*M*.
3. Water bath, 45°C.
4. 4-Vinylpyridine (4-VP) is distilled under reduced pressure (vacuum water pump). At the end of the distillation the apparatus is filled with inert gas (N_2 or Ar) to avoid oxidation of the colorless pure 4-VP. The reactive can be stored as pure liquid in sealed vials under inert gas and conserved for months at –20°C.
5. Nitrogen or Ar gas tank.

3. Methods

This chapter has been organized in a way that the reader can start from animals and end with structural information about the antimicrobial peptides present in the hemolymph or in the total body of the organism they are studying. We have chosen to describe the procedures in the order of normal execution, from induction of the antimicrobial peptides in the animal to sequencing and mass spectrometry.

3.1. Insect Immunization

1. Choose insects at the same developmental stage or animals of similar size.
2. Prepare overnight cultures of *Micrococcus luteus* (Gram-positive strain) and *Escherichia coli* wild-type (Gram-negative strain).
3. Anesthesize the animals by chilling or with CO_2.
4. Inject a mixture of living *M. luteus* and *E. coli* (1000 cells of each/µL injected) or, for small-size insects, replace the injection by simple pricking of individuals with a fine stainless steel needle previously dipped into a moist combined bacterial pellet.
5. Keep the bacteria-challenged insects in appropriate conditions for 24–48 h (*see* **Note 1**).

3.2. Extraction

3.2.1. Extraction from Hemolymph

1. Collect the hemolymph (through an incision, a leg, or an antenna) in precooled polypropylene tubes containing a protease inhibitor (for example aprotinin) at a final concentration of 10 µg/mL of hemolymph and an inhibitor of melanization (phenylthiourea, PTU, 1 µg/mL; *see* **Note 2**).
2. Recover cell-free hemolymph after centrifugation at 15,000*g* at 4°C for a short period to avoid coagulation.
3. Acidify the cell-free hemolymph with a solution of 0.1% trifluoroacetic acid (v:v) and, after incubation for 30 min in an ice-cold water bath under gentle shaking, centrifuge at 15,000*g* for 30 min, and collect the supernatant.

3.2.2. Extraction from Small-Sized Insects

1. Freeze insects in liquid nitrogen and reduce to a fine powder in a mortar in the continuous presence of liquid nitrogen (*see* **Note 3**).
2. Transfer the powder to 10 vol (w/v) of acidified water (0.1% TFA) containing a protease inhibitor (aprotinin, final concentration 10 µg/mL of medium) and an inhibitor of melanization (PTU, 1 µg/mL).
3. Extract as for the hemolymph, 30 min in a ice-cold water bath.
4. Centrifuge the extract, clarify through 0.8-µm filters and immediately (without freezing) submit to purification.

3.3. Purification of Antimicrobial Peptides

3.3.1. Solid-Phase Extraction (SPE) on Sep-Pak C18 Cartridges

1. Prepare the cartridge by washing with methanol and equilibrate with acidified water (0.05% TFA). Usually one cartridge is sufficient for 2–3 g of total animals or 1–2 mL of hemolymph.
2. Load the sample after verification of the pH of the extract, which should be acidic, pH <4.0 (*see* **Note 4**).

3. Elute in a stepwise fashion with increasing concentrations of acetonitrile in acidified water (0.05% TFA): 10, 40, and 80%.
4. Concentrate all fractions in a vacuum centrifuge or by lyophilization for large fractions (>25 mL).
5. Reconstitute the fractions in HPLC water and keep at –20°C until use.

3.3.2. Reversed-Phase HPLC

1. Load the sample on an analytical reversed-phase column (300 Å, 3.8–4.6 mm id, C18 or C8) equilibrated in 2% acetonitrile in acidified water (0.05% TFA).
2. Elute with an appropriate linear gradient of acetonitrile in acidified water (0.05% TFA; *see* **Note 5**) at a flow rate of 1 mL/min (*see* **Note 6**).
   a. For the 10% SPE fraction: 2–20% acetonitrile in 60 min.
   b. For the 40% SPE fraction: 2–60% acetonitrile in 120 min.
   c. For the 80% SPE fraction: 10–80% acetonitrile in 120 min.
3. Collect fractions manually according to the absorbance measured at 225 nm (best ratio signal/solvent). Using this collecting procedure, each fraction corresponds to an individual peak (*see* **Note 7**).
4. Dry the eluted fractions in vacuum centrifuge, dissolve in sterile distilled water, and monitor the antimicrobial activity on aliquots.

3.3.3. Size-Exclusion HPLC

1. Load the active fraction on two serially linked Beckman SEC 3000 and SEC 2000 columns, 300 × 7.5 mm (or equivalent columns) protected by a precolumn. The injected volume should be <100 μL for adequate resolution.
2. Elute with 30% acetonitrile in acidified water (TFA 0.05%) at a flow rate not exceeding 0.5 mL/min (*see* **Note 8**). Fractions are collected manually following the absorbance at 225 nm.
3. Dry the eluted fractions in a vacuum centrifuge, dissolve in sterile distilled water, and monitor the antimicrobial activity on aliquots.

3.3.4. Final Purification

1. Use a narrow-bore, reversed-phase column (approx 2 mm id).
2. Evaluate from the chromatogram obtained in **Subheading 3.3.2.** the exact concentration (C%) of acetonitrile for the elution of the compound of interest.
3. Perform the final chromatography using a linear diphasic gradient of acetonitrile in acidified water (TFA 0.05%) at a flow rate of 0.2–0.25 mL/min. The following two-step gradient can be used: 0% to ([C%] –2)% acetonitrile in 10 min and ([C%]–2)% to ([C%]+5)% acetonitrile in 40 min.
4. Collect the fractions by hand following the absorbance at 225 nm.

3.4. Microsequencing Analysis

1. Solubilize the sample (100–200 pmol) in a small volume (<10–15 μL) with a solution of 30% acetonitrile in acidified water (0.05% TFA).

2. Load the pure peptide in aliquots of 5 μL on an appropriate sequencing membrane. Dry the sample carefully for 20–30 min (*see* **Note 9**).

3.5. Mass Spectrometry

3.5.1. Electrospray Ionization Mass Spectrometry (ESI-MS) (see *Note 10*)

1. Dissolve the aliquot to analyze (approx OD_{225} = 0.005) in a solution of water/methanol (50:50; v:v) containing 1% acetic acid.
2. Inject a fivefold diluted solution first and if necessary, the remaining sample can be injected (*see* **Note 11**).

3.5.2. Matrix-Assisted Laser Desorption/Ionization Time-of-Flight Mass Spectrometry (MALDI-TOF-MS)

1. Dissolve the purified peptide (1 μL, equivalent to OD_{225} = 0.002) in 2.5 μL of a solution of α-cyano-4-hydroxycinnamic acid matrix (7% in 50% acetonitrile 0.05% TFA, w:v).
2. Transfer 1 μL of the peptide matrix solution to a stainless-steel target and dry under gentle vacuum.
3. Wash the sample on the target with 1 μL of TFA 0.1%.
4. Introduce the target in the mass spectrometer.

3.6. Enzymatic Cleavage

1. Add the enzyme in the appropriate digestion buffer (recommended by the manufacturer) on the dry sample.
2. Overnight incubation at 37°C.
3. The reaction is stopped by direct separation of the reaction mixture on reversed-phase HPLC.
4. The collected fractions are dried in vacuum centrifuge.
5. The fragments are analyzed by mass spectrometry and/or sequencing (*see* **Note 12**).

3.7. Bioassays

3.7.1. Collection of Fungal Spores

1. Pour Petri dishes with six cereal agar (CA).
2. Transfer a mycelium plug of *Neurospora crassa* to the center of a 6 CA medium plate (*see* **Note 13**).
3. Seal with parafilm.
4. Grow at room temperature under white fluorescent light.
5. Check spore formation: Place mycelium plug in a drop of water on a slide and analyze under microscope.
6. Cover each Petri dish with 5–10 mL of sterile water. Rub the surface with sterilized spatula. Filter the suspension containing the mycelium and the spores over

sterilized glass wool-plugged funnel and collect the spore suspension in a sterilized polypropylene centrifuge tube.

7. Wash the spore suspension twice by centrifugation (15000*g*, 15 min) and resuspend in water.
8. Determine the spore density in a Thoma chamber or a related hematocytometric chamber.
9. Adjust the spore density to 2×10^7 spores per mL. Transfer 4 (or more) × 600 µL to sterilized microtubes. To each tube add 600 µL of 50% glycerol. Vortex each tube and transfer their contents in 100-µL aliquots in sterilized microtubes (*see* **Note 14**). Store at –80°C.

3.7.2. Antibacterial and Antifungal Assays

Antimicrobial (antibacterial and antifungal) activity is monitored during the different purification steps by liquid growth inhibition assays.

3.7.2.1. Antifungal Assay

1. Suspend spores in a growth medium containing potato dextrose broth (Difco, E. Molesley, Surrey, UK, in half-strength) supplemented with tetracycline (100 µg/mL) and cefotaxim (1 µg/mL).
2. Place 20 µL of either water (control) or 10 µL of the fractions supplemented with 10 µL of water into wells of a 96-well microtiter plate.
3. Add 80 µL of the spore suspension to each well.
4. Incubate plates 48 h at 25°C in the dark in a moist chamber (*see* **Note 15**).
5. Evaluate growth of fungi by measuring absorbance of the culture at 600 nm using a microplate reader (*see* **Note 16**).

3.7.2.2. Antibacterial Assay

This test is easy, sensitive, and can be used for a lot of fractions. Intermediary measurements can be made during incubation to detect molecules with lower activity. Using sequential dilution, the minimal growth inhibition concentration (MIC) can be easily determined. MIC is expressed as the lowest final concentration of the peptide at which no growth is observed ***(6,22)*** or, more precisely, MIC has been defined as an interval (a–b), being the highest concentration tested at which cells are able to grow, and the lowest concentration tested that inhibit bacterial growth completely ***(7–9)***.

1. From one colony, prepare an overnight culture of *M. luteus* (Gram-positive strain) and *E. coli* D31 (Gram-negative strain, a sensitive strain) in Luria Bertani's rich medium at 37°C.
2. Measure the absorbance at 600 nm and inoculate Poor Broth medium at 25°C to obtain a fresh exponential phase culture.

3. Dilute bacteria in Poor Broth medium at a starting $OD_{600} = 0.001$.
4. Add 10-µL aliquots of either water (control) or of the fraction to test (*see* **Note 17**).
5. Place 100 µL of the exponential phase culture at the starting $OD_{600} = 0.001$ in each well of a microplate.
6. Incubate plates with gentle shaking at 25°C.
7. Evaluate bacterial growth by measuring the absorbance of the culture at 600 nm using a microplate reader.

3.8. Reduction and Alkylation

During peptide sequencing, there is no signal when a cysteine residue appears in a sequence. However, cysteines can be detected by reduction of the disulfide bridge and alkylation of the sulfhydryl group before sequencing.

1. Dissolve the peptide (100 pmol to 1 nmol) in the reaction buffer.
2. Add 2 µL of 2.2*M* dithiothreitol.
3. Flush the sample with N_2 to prevent oxidation.
4. Incubate at 45°C for 1 h in the dark.
5. Add pure (colorless) or freshly distilled 4-vinylpyridine (2 µL; *see* **Note 18**).
6. Flush the sample with N_2 to prevent oxidation.
7. Incubate as in **step 4**, but for 10 min.
8. Stop the reaction by injection of the reactive mixture on a reversed-phase columns. The pyridylethylated peptide is separated from reagents and unreacted peptide by a gradient of acetonitrile in acidified water from 2 to 80% on 120 min at 1 mL/min.

4. Notes

1. A control experiment with unchallenged insects is important to confirm the inducibility of the molecules.
2. An equivalent of 2–3 mL of cell-free hemolymph is often required to obtain structural information about the antimicrobial peptides.
3. An equivalent of 2–10 g of individuals is often necessary to get structural information about antimicrobial peptides.
4. When several cartridges are used for the same sample, loading is performed on serially connected columns while each cartridge is treated individually for elution.
5. Hydrochloric acid can also be used to acidify the medium, for enhanced chromatographic selectivity, but in this case an inert (nonmetallic) system is suitable *(32)*.
6. To facilitate the separation, if your HPLC is equipped with an oven, the temperature of the column can be increased (should not exceed 40°C).
7. The elution of peptides from the column is monitored by measuring the UV absorbance at 280 nm (detection of aromatic amino acids, low sensitivity) or between

214 and 225 nm (detection of the peptide bond, high sensitivity but absorption of solvent at this low wavelength is strong). It is preferable to collect the peaks by hand rather than using an automatic collector to minimize the number of purification steps. This collecting is highly facilitated by the use of an analogic direct recorder. Each fraction is concentrated with vacuum centrifuge and an aliquot is then tested for antimicrobial activity.

8. The size-exclusion column should be equilibrated in 30% acetonitrile in acidified water (TFA 0.05%) during several hours for a good stabilization of the phase. The column should be stored in 30% acetonitrile without acid. Do not exceed 30% acetonitrile on these columns.
9. Usually, 100 pmol of sample are sufficient to get a sequence of 30–40 residues. Never discard the sequencing membrane if no sequence is obtained. Proteolytic or chemical cleavage and separation of the peptidic fragments can be done directly on the membrane.

   During sequencing, if no phenylthiohydantoin (PTH) amino acid can be detected in some cycles or if peaks corresponding to an unidentified residue are found, this indicates that there is either a cysteine that is not detected during Edman degradation or that the amino acid is modified by a posttranslational modification ***(18,21,29,33)***. The total absence of sequence can result from the N-terminal blocking of the sequence (acetylation, formylation, or presence of a pyroglutamate).
10. The choice of the MS technique, when it is possible, will be directed by the sample. When pure and a picomole level of molecules is available, ESI-MS is recommended. For a mixture or for very low quantities of material, MALDI-TOF-MS will be better.

    ESI-MS has superior mass accuracy compared to MALDI-TOF-MS, whereas the latter is considered to have higher tolerance to mixture and solvent conditions.
11. An external mass calibration is provided with peptides with a molecular weight close to the expected masses.
12. Enzymatic cleavages are more efficient on reduced and alkylated peptides when cysteines are present in the molecule.
13. To get the first *Neurospora crassa* spore preparation, you have to culture some of the spores (10 µL) in a Petri dish with 6 CA.
14. With one tube of 100 µL you can test more than 800 fractions.
15. A control of growth delay can be monitored after 24 h using microscope evaluation.
16. *Neurospora crassa* (a filamentous fungus) is very sensitive and is a good strain to use for such studies.
17. An equivalent of 150 mg of total insect is recommended for measuring antimicrobial activity in a liquid growth inhibition assay. If the extract is prepared from hemolymph, a fraction aliquot corresponding to 5–20 µL of hemolymph is sufficient to detect antimicrobial activities. The following scheme (**Fig. 1**) represents a classical organization of a liquid growth inhibition assay.

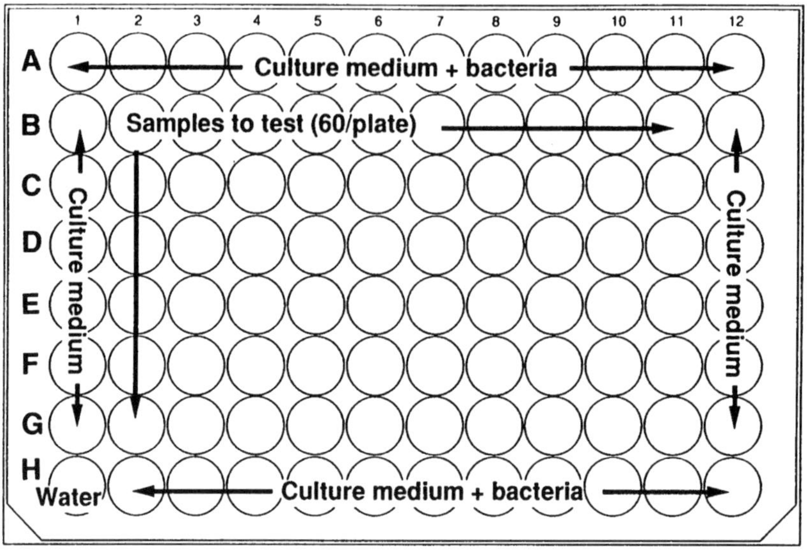

Fig. 1. Position of the samples in a microtitration plate. The first and the last lines contain culture medium with bacteria. Well H1 contains 100 μL of water as a blank. Columns B1 to G1 and B12 to G12 contain culture medium without bacteria and can serve as control for accidental contamination of the medium. The central wells of lines B to G (60 per plate) are used to analyze the samples. Positive controls, with known antibacterial molecules are currently put in wells H10 and H11. If there are not enough samples to complete the plate, empty wells are filled with water or inoculated medium so that the plates are always incubated under the same conditions.

18. The 4-vinylpyridine should be colorless and this protection is preferred to other protecting SH groups because the PTH-Cys 4VP is easily available and allows its identification during sequencing. Distillation of 4-VP is recommended in a chemically well-equiped laboratory, under a hood, and glasses must be worn.

References

1. Hetru, C., Hoffmann, D., and Bulet, P. (1997) Antimicrobial peptides from insects, in *Molecular Mechanisms of Immune Responses in Insects* (Brey, P. T. and Hultmark, D., eds.) Chapman & Hall, London, in press.
2. Steiner, H., Hultmark, D., Engström, A., Bennich, H., and Boman, H.G. (1981) Sequence and specificity of two antibacterial proteins involved in insect immunity. *Nature* **292,** 246–248.
3. Hetru, C., Bulet, P., Cociancich, S., Dimarcq, J.-L., Hoffmann, D., and Hoffmann, J.A. (1994) Antibacterial peptides/polypeptides in the insect host defense: a comparison with vertebrate antibacterial peptides/polypeptides, in *Phylogenetic Per-*

spectives in Immunity: The Insect Host Defense (Hoffmann, J. A., Janeway, Jr, C. A., and Natori, S., eds.) Landes Company, Austin, pp. 43–66.

4. Cociancich, S., Bulet, P., Hetru, C., and Hoffmann, J. A. (1994) The inducible antibacterial peptides of insects. *Parasitology Today* **10,** 132–139.
5. Qu, X.-M., Steiner, H., Engström, A., Bennich, H., and Boman, H.G. (1982) Insect immunity: isolation and structure of cecropins B and D from pupae of the chinese oak silk moth, *Antheraea pernyi. Eur. J. Biochem.* **127,** 219–224.
6. Casteels, P., Ampe, C., Jacobs, F., Vaeck, M., and Tempst, P. (1989) Apidaecins: antibacterial peptides from honeybees. *EMBO J.* **8,** 2387–2391.
7. Casteels, P., Ampe, C., Riviere, L., Van Damme, J., Elicone, C., Fleming, M., Jacobs, F., and Tempst, P. (1990) Isolation and characterization of abaecin, a major antibacterial peptide in the honeybee (*Apis mellifera*). *Eur. J. Biochem.* **187,** 381–386.
8. Casteels, P., Romagnolo, J., Castle, M., Casteels-Josson, K., Erdjument, H., and Tempst, P. (1994) Biodiversity of apidaecin-type antibiotics. *J. Biol. Chem.* **269,** 26,107–26,115.
9. Casteels, P., Ampe, C., Jacobs, F., and Tempst, P. (1993) Functional and chemical characterization of hymenoptaecin, an antibacterial polypeptide that is infection-inducible in the honeybee (*Apis mellifera*). *J. Biol. Chem.* **268,** 7044–7054.
10. Ourth, D. D., Lockey, T. D., and Renis, H. E. (1994) Induction of cecropin-like and attacin-like antibacterial but not antiviral activity in *Heliothis virescens* larvae. *Biochem. Biophys. Res. Commun.* **200,** 35–44.
11. Lowenberger, C., Bulet, P., Charlet, M., Hetru, C., Hodgeman, B., Christensen, B. M., and Hoffmann, J. A. (1995) Insect immunity: isolation of three novel inducible antibacterial defensins from the vector mosquito, *Aedes aegypti. Insect Biochem. Mol. Biol.* **25,** 867–873.
12. Dickinson, L., Russel, V., and Dunn, P. E. (1988) A family of bacteria-regulated cecropin D-like peptides from *Manduca sexta. J. Biol. Chem.* **263,** 19,424–19,429.
13. Boucias, D. G., Hung, S. Y., Mazet, I., and Azbell, J. (1994) Effect of the fungal pathogen, *Beauveria bassiana*, on the lysozyme activity in *Spodoptera exigua* larvae. *J. Insect Physiol.* **40,** 385–391.
14. Morishima, I., Horiba, T., and Yamano, Y. (1994) Lysozyme activity in immunized and non-immunized hemolymph during the development of the silkworm, *Bombyx mori. Comp. Biochem. Physiol.* **108A,** 311–314.
15. Lambert, J., Keppi, E., Dimarcq, J.-L., Wicker, C., Reichhart, J.-M., Dunbar, B., Lepage, P., Van Dorsselaer, A., Hoffmann, J. A., Fothergill, J., and Hoffmann, D. (1989) Insect immunity: isolation from immune blood of the dipteran *Phormia terranovae* of two insect antibacterial peptides with sequence homology to rabbit lung macrophage bactericidal peptides. *Proc. Natl Acad. Sci. USA*, **86,** 262–266.
16. Bulet, P., Cociancich, S., Dimarcq, J.-L., Lambert, J., Reichhart, J.-M., Hoffmann, D., Hetru, C., and Hoffmann, J. A. (1991) Isolation from a coleopteran insect of a novel inducible antibacterial peptide and of new members of the insect defensin family. *J. Biol. Chem.* **266,** 24,520–24,525.

17. Bulet, P., Dimarcq, J.-L., Hetru, C., Lagueux, M., Charlet, M., Hégy, G., Van Dorsselaer, A., and Hoffmann, J. A. (1993) A novel inducible antibacterial peptide of *Drosophila* carries an O-glycosylated substitution. *J. Biol. Chem.* **268,** 14,893–14,897.
18. Bulet, P., Cociancich, S., Reuland, M., Sauber, F., Bischoff, R., Hégy, G., Van Dorsselaer, A., Hetru, C., and Hoffmann, J. A. (1992) A novel insect defensin mediates the inducible antibacterial activity in larvae of the dragonfly *Aeschna cyanea* (Paleoptera, Odonata). *Eur. J. Biochem.* **209,** 977–984.
19. Cociancich, S., Goyffon, M., Bontems, F., Bulet, P., Bouet, F., Menez, A., and Hoffmann, J. A. (1993) Purification and characterization of a scorpion defensin, a 4 kDa antibacterial peptide presenting structural similarities with insect defensins and scorpion toxins. *Biochem. Biophys. Res. Commun.* **194,** 17–22.
20. Cociancich, S., Dupont, A., Hégy, G., Lanot, R., Holder, F., Hetru, C., Hoffmann, J. A., and Bulet, P. (1994) Novel inducible antibacterial peptides from a hemipteran insect, *Pyrrhocoris apterus. Biochem. J.* **300,** 567–575.
21. Dimarcq, J.-L., Hoffmann, D., Meister, M., Bulet, P., Lanot, R., Reichhart, J.-M., and Hoffmann, J. A. (1994) Characterization and transcriptional profiles of a *Drosophila* gene encoding an insect defensin. *Eur. J. Biochem.* **221,** 201–209.
22. Nakamura, T., Furunaka, H., Miyata, T., Tokunaga, F., Muta, T., Iwanaga, S., Niva, M., Takao, T., and Shimonishi, Y. (1988) Tachyplesin, a class of antimicrobial peptide from the hemocytes of the horseshoe crab (*Tachypleus tridentatus*) *J. Biol. Chem.* **263,** 16709–16713.
23. Saito, T., Kawabata, S.-I., Shigenaga, T., Takayenski, Y., Cho, J., Nakajima, H., Hirata, M., and Iwanaga, S. (1995) A novel big defensin identified in horseshoe crab hemocytes: isolation, amino acid sequence, and antibacterial activity. *J. Biochem.* **117,** 1131–1137.
24. Matsuyama, K. and Natori, S. (1988) Purification of three antibacterial proteins from the culture medium of NIH Sape-4, an embryonic cell line of *Sarcophaga peregrina. J. Biol. Chem.* **263,** 17,112–17,116.
25. Yamada, K. and Natori, S. (1993) Purification, sequence and antibacterial activity of two novel sapecin homologues from *Sarcophaga* embryonic cells: similarity of sapecin B to charybdotoxin. *Biochem. J.* **291,** 275–279.
26. Marchini, D., Giordano, P. C., Amons, R., Bernini, L. F., and Dallai, R. (1993) Purification and primary structure of ceratotoxin A and B, two antibacterial peptides from the female reproductive accessory glands of the medfly *Ceratitis capitata* (Insecta: Diptera). *Insect Biochem. Mol. Biol.* **23,** 591–598.
27. Lee, W.-J. and Brey, P. T. (1994) Isolation and identification of cecropin antibacterial peptides from the extracellular matrix of the insect integument. *Anal. Biochem.* **217,** 231–235.
28. Hara, S. and Yamakawa, M. (1995) A novel antibacterial peptide family isolated from the silkworm *Bombyx mori. Biochem. J.* **310,** 651–656.
29. Hultmark, D., Engström, A., Bennich, H., Kapur, R., and Boman, H. G. (1982) Insect immunity: isolation and structure of cecropin D and four minor antibacterial components from *cecropia* pupae. *Eur. J. Biochem.* **127,** 207–217.

30. Abraham, E. G., Nagaraiu, J., Salunke, D., Gupta, H. M., and Datta, R. K. (1995) Purification and partial characterization of an induced antibacterial protein in the silkworm *Bombyx mori. J. Invert. Pathol.* **65,** 17–24.
31. Morishima, I., Suginaka, S., Ueno, T., and Hirano, H. (1990) Isolation and structure of cecropins, inducible antibacterial peptides, from the silkworm, *Bombyx mori. Comp. Biochem. Physiol.* **958,** 551–554.
32. Young P. M. and Wheat T. E. (1990) Alternative mobile phase for enhanced chromatographic selectivity and increased sensitivity in peptide separations. *Pept. Res.* **3,** 287–292.
33. Bulet, P., Hégy, G., Lambert, J., Van Dorsselaer, A., Hoffmann, J. A., and Hetru, C. (1995) Insect immunity. The inducible antibacterial peptide diptericin carries two O-glycans necessary for biological activity. *Biochemistry.* **34,** 7394–7400.

4

Big Defensin and Tachylectins-1 and -2

Shun-ichiro Kawabata and Sadaaki Iwanaga

1. Introduction

Hemocytes of the horseshoe crab contain a family of arthropod peptide antibiotics, termed the tachyplesin family, and an antibacterial protein, called anti-LPS factor, of which the former is located in the small (S) granules and the latter in the large (L) granules of the hemocytes *(1–5)*. In our ongoing studies on granular components, we have identified a novel defensin-like substance, named big defensin, present in both L- and S-granules *(6)*. This substance strongly inhibits the growth of Gram-negative and Gram-positive bacteria, and fungi, such as *Candida albicans* (**Table 1**). The isolated big defensin consists of a total of 79 amino acid residues, in which the COOH-terminal region composed of 37 amino acids resembles mammalian defensins. Big defensin, however, is distinct from the mammalian defensins in molecular size, the latter of which have 29–34 amino acid residues in common *(7,8)*. The disulfide motif in the defensin-like domain of big defensin is identical to that of ß-defensin from bovine neutrophils but not to that of classical defensins *(9,10)*. Furthermore, the structural organization of big defensin differs markedly from those of insect defensins not only in disulfide bridge locations but also in the molecular size *(11,12)*.

Big defensin shows a highly amphipathic nature, a high hydrophobicity in the NH_2-terminal region, and a hydrophilicity with clustering of cationic residues in the COOH-terminal region. A noteworthy characteristic of big defensin is that there is a functional difference between the NH_2-terminal portion corresponding to residues 1–37 and the COOH-terminal defensin portion corresponding to residues 38–79. Although intact big defensin has antibacterial activities against both *Salmonella typhimurium* and *Staphylococcus aureus*,

From: *Methods in Molecular Biology, Vol. 78: Antibacterial Peptide Protocols*
Edited by: W. M. Shafer, Humana Press Inc., Totowa, NJ

Table 1
Antimicrobial Activities of Big Defensin and Tachylectin-1

	IC_{50} (μg/mL)	
	Big Defensin	Tachylectin-1
Gram-negative bacteria		
Escherichia coli 09:K39 (K^-)	2.5	10–20
Salmonella typhimurium LT2 (S)	<1.3	ND
Salmonella minnesota R595 (Re)	<0.6	20–40
Klebsiella pneumoniae	<1.3	20–40
Gram-positive bacteria		
Staphylococcus aureus	<2.5	ND
Fungus		
Candida albicans	<10	

IC_{50}, 50% inhibitory concentration.
ND, not detectable at 40 μg/mL.

the NH_2-terminal hydrophobic fragment is more effective than the COOH-terminal defensin fragment against Gram-positive bacteria. In contrast, the COOH-terminal defensin fragment displays a more potent activity than the NH_2-terminal hydrophobic fragment against Gram-negative bacteria.

Big defensin has no hemolytic activity against sheep erythrocytes, but it shows the erythrocyte-agglutinating activity, so-called LPS-binding activity, using sheep erythrocytes sensitized with *Salmonella minnesota* Re-LPS (**Table 2**). However, the NH_2- and COOH-terminal fragments, in addition to the S-alkylated big defensin, have weak LPS-binding activity, indicating that the native conformation of the entire molecule is required for its binding with LPS.

A lectin-like substance, tachylectin-1 (L6), in the large granule of hemocytes, has recently been purified by affinity chromatography of LPS immobilized-agarose ***(13)***. Tachylectin-1 shows antibacterial activity toward Gram-negative but not Gram-positive bacteria (**Table 1**). It also has more effective agglutinating activity toward Gram-negative than Gram-positive bacteria (**Table 3**). Tachylectin-1 binds also to agarose itself and it is eluted with high concentrations of monosaccharides of 0.5–1*M*, such as glucose, mannose, and galactose. On the other hand, the purified tachylectin-1 displays LPS-binding potential, agglutinates sheep erythrocytes coated with LPS, and activity is inhibited by addition of free LPS. However, it apparently has no hemagglutinin activity for sheep and rabbit red blood cells, and human A, B, and O types of red blood cells. These results would suggest that tachylectin-1 recognizes an oligosaccharide portion of LPS. Inductively coupled plasma spectrometry suggests the presence of 0.75 mol zinc/mol protein. EDTA or *o*-phenanthroline is

Table 2
LPS-Binding Activities of Big Defensin and Its Derivatives

	MAC (μg/mL)
	LPS-Binding Activity
Big defensin	6.3
NH_2-terminal fragment	50
COOH-terminal fragment	ND
S-Alkylated big defensin	25

MAC, minimum agglutinating concentration of *S. minnesota* Re. LPS-sensitized sheep erythrocytes.
ND, not detectable at 40 μg/mL.

also an effective eluant for tachylectin-1 bound with LPS-agarose and therefore, zinc ions must have an important role for the sugar-binding ability of tachylectin-1.

Tachylectin-1 is a single-chain protein consisting of 221 amino acids with no N-linked sugar chain and contains three intrachain disulfide bonds and a free Cys residue. An outstanding structural feature of tachylectin-1 is that it consists of six tandem repeats, each containing 33–38 amino acids with 32–61% internal sequence identities. Tachylectin-1 shows no significant sequence similarity with other known proteins, including various animal and plant lectins and LPS-binding proteins.

Tachylectin-2 (L10), an L-granule component, has been proved to be a lectin with hemagglutinating activity against human A-type erythrocytes and with specificity for GlcNAc. At least three isoproteins of tachylectin-2 (2a, 2b, and 2c) have been isolated ***(14)***. Several sugars containing the *N*-acetyl group inhibit the hemagglutinating activity of tachylectin-2, but *N*-acetylneuraminic acid and *N*-acetylmuramic acid do not. In addition, *N*-acetylallolactosamine (Galß1-6GlcNAc) has the inhibitory activity but its isomer, *N*-acetyllactosamine (Galß1-4GlcNAc) is without effect. Furthermore, tachylectin-2 has an agglutination activity against *Staphylococcus saprophyticus* KD (**Table 3**) and the agglutination is inhibited in the presence of GlcNAc.

Tachylectin-2 is a single-chain protein consisting of 236 amino acids with no cysteine and no Asn-linked and O-linked sugar chains, and it has no significant sequence similarity with other known proteins, including various animal lectins. The most intriguing feature of the amino acid sequence of tachylectin-2 is the presence of an internal homology that consists of five tandem repeats of 47 amino acids. Tachylectin-2 is present as a monomeric protein in solution, as deduced from data in the ultracentrifugal analysis. This is also a unique

Table 3
Bacterial Agglutinating Activities of Tachylectin-1 and Tachylectin-2

	MAC (μg/mL)	
	Tachylectin-1	Tachylectin-2
Gram-negative bacteria		
Escherichia coli K12	25	ND
Escherichia coli B	12.5	ND
Gram-positive bacteria		
Staphylococcus saprophyticus	200	62.5
Micrococcus luteus	200	ND
Enterococcus hirae	50	ND

MAC, minimum agglutinating concentration.
ND, not detectable at 500 μg/mL.

characteristic of tachylectin-2, compared to other horseshoe crab agglutinins so far purified, such as, limulin from *Limulus polyphemus* ***(15)***, carcinoscorpin from *Carcinoscorpius rotundicauda* ***(16)***, and sialic acid-specific binding lectin from *Tachypleus tridentatus* ***(17)***. These lectins are present in hemolymph as high molecular mass oligomers of 460, 420, and 533 kDa, respectively.

2. Materials

2.1. Preparation of Horseshoe Crab Hemocyte Lysate and Debris

1. Refer to Chapter 5.

2.2. Purification of Big Defensin

1. 30% (v/v) acetic acid.
2. Sephadex G-50 (from Pharmacia, Uppsala, Sweden).
3. 10% (v/v) acetic acid.
4. S-Sepharose FF (Pharmacia).
5. 20 m*M* Tris-HCl, 0.1*M* NaCl, pH 8.0.
6. 20 m*M* Tris-HCl, 0.4*M* NaCl, pH 8.0.

2.3. Purification of Tachylectin-1

2.3.1. Preparation of LPS-Immobilized Agarose

1. LPS (*Escherichia coli* O111: B4, List Biological Laboratories, Campbell, CA).
2. Epoxy-activated Sepharose CL-6B FF (Pharmacia).
3. 0.1*M* sodium carbonate, pH 11.0.
4. 0.1*M* sodium bicarbonate, pH 8.0.

5. 0.1*M* sodium acetate, pH 4.0.
6. 1.0*M* ethanol amine.

2.3.2. Preparation of Lubrol Extract

1. 10% Lubrol PX.
2. 20 m*M* Tris-HCl, 0.15*M* NaCl, pH 7.5.

2.3.3. LPS-Agarose Column Chromatography

1. 20 m*M* Tris-HCl, 0.15*M* NaCl, pH 7.5.
2. 20 m*M* Tris-HCl, 0.15*M* NaCl, 1*M GlcNAc*, pH 7.5.

2.4. Purification of Tachylectin-2

2.4.1. Preparation of Dextran Sulfate-Sepharose CL-6B (see Chapter 5)

2.4.2. Dextran Sulfate-Sepharose CL-6B Column Chromatography (see Chapter 5)

2.4.3. CM-Sepharose CL-6B Column Chromatography

1. CM-Sepharose CL-6B (Pharmacia).
2. 20 m*M* Tris-HCl, pH 7.5.
3. 20 m*M* Tris-HCl, 0.15*M* NaCl, pH 7.5.

2.4.4. Mono S Column Chromatography

1. Mono S HR 5/5 (Pharmacia).
2. 50 m*M* Sodium acetate, pH 5.5.
3. 50 m*M* Sodium acetate, 0.2*M* NaCl, pH 5.5.

2.5. Antimicrobial Activity

1. Bacteria: *Salmonella typhimurium* LT2 (smooth), *Salmonella minnesota* R595 (Re mutant), *Escherichia coli* 09: K39 (K^- strains), *Klebsiella pneumoniae*, *Staphylococcus aureus*, and *Candida albicans* (*see* **Note 1**).
2. 3% (w/v) Tryptosoy broth sterilized by autoclaving for 20 min (*see* **Note 5**).
3. 10 m*M* sodium phosphate, pH 7.0.
4. 1% (w/v) Agar plates containing 3% Tryptosoy broth.

2.6. LPS-Binding Assay

1. Sheep defibrinated blood (*see* **Note 6**).
2. 10 m*M* sodium phosphate, 0.15*M* NaCl, pH 7.0.
3. LPS (100 µg/mL, *S. minnesota* R595) in 10 m*M* sodium phosphate, 0.15*M* NaCl, pH 7.0.
4. U-bottomed microtiter plates.

2.7. Bacterial Agglutination Assay

1. Bacteria: *Staphylococcus aureus* 209 P, *Staphylococcus epidermidis* K3, *Staphylococcus saprophyticus* KD, *Micrococcus luteus*, *Enterococcus hirae*, and *Escherichia coli* strain B.
2. 3% Tryptosoy broth sterilized by autoclaving for 20 min.
3. 20 m*M* Tris-HCl, 0.15*M* NaCl, pH 7.5.
4. U-bottomed microtiter plates.

2.8. Hemagglutination Assay

1. Human A-, B-, or O-type erythrocytes (out-dated concentrated erythrocytes) (*see* **Note 9**).
2. 20 m*M* Tris-HCl, 0.15*M* NaCl, pH 7.5.
3. 0.5*M* $CaCl_2$.
4. U-bottomed microtiter plates.

3. Methods

3.1. Preparation of Horseshoe Crab Hemocyte Lysate and Debris (see Chapter 5)

3.2. Purification of Big Defensin

3.2.1. Acid Extraction of Debris

1. Suspend 30 g of the hemocyte debris in 200 mL of 30% acetic acid.
2. Homogenize the suspension with Physcotron® for 1 min and stir for 30 min at 4°C.
4. Centrifuge at 14,000*g* for 15 min at 4°C. Transfer the supernatant into a glass container.
5. Re-extract the pellet twice more with 200 mL of 30% acetic acid. Collect the supernatant (600 mL in total) and dilute with 1.2 L of distilled water.
6. Lyophilize the acid extract.

3.2.2. Sephadex G-50 Column Chromatography

1. Dissolve the lyophilized material in 50 mL of 10% acetic acid.
2. Apply the sample to a Sephadex G-50 column (3.6 × 110 cm) equilibrated with 10% acetic acid.
3. Perform SDS-PAGE in 15% gel of every five tubes.
4. Pool fractions containing the 8-kDa protein.
5. Lyophilize the pooled fractions.

3.2.3. S-Sepharose FF Column Chromatography

1. Dissolve the lyophilized protein in 100 mL of 20 m*M* Tris-HCl, 0.1*M* NaCl, pH 8.0, and centrifuge at 4000*g* for 5 min to remove insoluble material.
2. Apply the supernatant to an *S*-Sepharose FF column (2 × 32 cm), equilibrated with the same buffer.

3. Wash the column extensively with the same buffer.
4. Elute proteins with a linear gradient (600 ml) of 0.1–0.4*M* NaCl in the same buffer. Two separated peaks are obtained, both contain proteins with an 8-kDa band on SDS-PAGE in 15% (w/v) gel.
5. Pool the second peak containing big defensin.

3.3. Purification of Tachylectin-1

3.3.1. Preparation of LPS-Immobilized Agarose

1. Suspend 5 mg of LPS in 20 mL of 0.1*M* sodium carbonate (pH 11.0) and sonicate in a ultrasonic bath (Bransonic, Model B1200J-1, Danbury, CT) for 10 min.
2. Wash epoxy-activated Sepharose CL-6B FF (3.3 g) with 400 mL of distilled water on a glass filter.
3. Mix the gel with the LPS suspension and rock for 16 h at room temperature.
4. Wash the gel with distilled water, 0.1*M* sodium bicarbonate (pH 8.0), and 0.1*M* sodium acetate (pH 4.0; 100 mL of each).
5. Suspend the gel in 1*M* ethanol amine to block the remaining functional group.

3.3.2. Preparation of Lubrol Extract

1. Homogenize 20 g of the debris in a Waring blender for 3 min in 500 mL of 20 m*M* Tris-HCl, 0.15*M* NaCl, pH 7.5.
2. Add 10% Lubrol PX to the homogenate to give a final concentration of 0.5%.
3. Stir the mixture for 3 h at 4°C.
4. Dialyze for overnight against 20 m*M* Tris-HCl, 0.15*M* NaCl, pH 7.5.
5. Centrifuge at 14,000*g* for 30 min at 4°C and transfer the supernatant into a glass container.

3.3.3. LPS-Agarose Column Chromatography

1. Apply the Lubrol extract to an LPS-agarose column (10 mL), equilibrated with 20 m*M* Tris-HCl, 0.15*M* NaCl, pH 7.5.
2. Wash the column extensively with the equilibration buffer.
3. Elute tachylectin-1 with the same buffer containing 1*M* GlcNAc.

3.4. Purification of Tachylectin-2

3.4.1. Preparation of Dextran Sulfate-Sepharose CL-6B

1. Prepare lysate from 50 g of hemocytes. For details, refer to the section of tachyplesin and anti-LPS factor of this volume.

3.4.2. Dextran Sulfate-Sepharose CL-6B Column Chromatography

1. Apply the hemocyte lysate to a dextran sulfate-Sepharose CL-6B (5 × 23 cm) as described in **Subheading 3.2.2.** of Chapter 5.
2. Perform SDS-PAGE in 12% gel on every 10 tubes.

4. Pool fractions containing the 27-kDa protein in the flow-through fraction, especially in the shoulder of the fractions.

3.4.3. CM-Sepharose CL-6B Column Chromatography

1. Apply the pooled fraction to CM-Sepharose CL-6B (2.1 × 12 cm), equilibrated with 20 m*M* Tris-HCl (pH 7.5).
2. Wash the column extensively with the equilibration buffer.
3. Elute tachylectin-2 with a linear gradient (400 mL) of 0–0.15*M* NaCl in the same buffer and pool the eluted fractions containing the 27-kDa protein.
4. Dialyze the pooled sample against 50 m*M* sodium acetate (pH 5.5).

3.4.4. Mono S Column Chromatography

1. Apply the dialyzed sample to Mono S HR 5/5, equilibrated with 50 m*M* sodium acetate (pH 5.5).
2. Wash the column with the equilibration buffer.
3. Elute tachylectin-2 isoforms in order, tachylectin-2a, -2b, and -2c, with a linear gradient of 0–0.2*M* NaCl in the same buffer.

3.5. Antimicrobial Activity

1. Culture bacteria in 3 mL of 3% Tryptosoy broth for 12 h at 37°C.
2. Collect bacteria by centrifugation at 4000*g* for 2 min and wash them with 10 m*M* sodium phosphate (pH 7.0).
3. Suspend the washed bacteria in 3 mL of 10 m*M* sodium phosphate (pH 7.0) and dilute 10^5-fold with the same buffer (5×10^3 to 1×10^4 cells/mL) (*see* **Notes 2** and **3**).
4. Add 50 μL of a solution of antimicrobial proteins or lectins to 450 μL of the bacterial suspension. As a control experiment, add the phosphate buffer to the bacteria suspension.
5. Incubate the mixture for 1 h at 37°C.
6. Plate 100 μL of the reaction mixture onto 1% agar plates containing 3% Tryptosoy broth (use a guanofracin-Sabouraud agar plate for *Candida albicans*).
7. Count the number of colonies on the plates after 24 h of incubation at 37°C and make the number 10 times to obtain colony forming units (CFU, number of colonies/mL).
8. Antimicrobial activity is usually expressed as IC_{50}, 50% inhibitory concentration of antimicrobial substances (*see* **Note 4**).

3.6. LPS-Binding Assay

1. Centrifuge sheep defibrinated blood at 1000*g* for 3 min and wash erythrocytes three times with 10 m*M* sodium phosphate, 0.15*M* NaCl, pH 7.0.
2. Prepare 1% erythrocytes suspension (vol/vol) in the same buffer.
3. Suspend 1 mg of LPS in 10 mL of 10 m*M* sodium phosphate, 0.15*M* NaCl, pH 7.0, and sonicate in an ultrasonic bath for 10 min.

4. Mix 0.2 mL of the LPS solution with 1 mL of the erythrocytes suspension and incubate for 30 min at 37°C.
5. Wash the erythrocytes with 10 m*M* sodium phosphate, 0.15*M* NaCl, pH 7.0, and prepare 1% suspension of LPS-sensitized erythrocytes in the same buffer.
6. Mix 50 μL of the LPS-sensitized erythrocytes with 50 μL of a twofold serial dilution of antimicrobial peptides or lectins in a U-bottomed microtiter plate and incubate for 1 h at 37°C.
7. The LPS-binding activity is expressed as the minimum agglutinating concentration of the sample tested.
8. For a competition assay, premix 25 μL of an LPS solution and 25 μL of samples before adding the LPS-sensitized erythrocytes. Inhibition of the free LPS is expressed as the minimum inhibitory concentration of LPS.

3.7. Bacterial Agglutination Assay

1. Culture bacteria in 5 mL of 3% Tryptosoy broth for 12 h at 37°C.
2. Collect bacteria by centrifugation at 4000*g* for 2 min and wash them with 20 m*M* Tris-HCl, 0.15*M* NaCl, pH 7.5.
3. Suspend the washed bacteria in the same buffer to give an absorbance at 600 nm of 10 (add about 1.0 mL of the buffer) (*see* **Note 8**).
4. Mix 25 μL of the suspension of each bacteria with 25 μL of a twofold serial dilution of antimicrobial peptides or lectins in a U-bottomed microtiter plate and incubate at room temperature for overnight.
5. The agglutinating activity is expressed as the minimum agglutinating concentration of the sample tested.
6. For screening of inhibitors, premix 12.5 μL of an inhibitor and 12.5 μL of the lectins before adding the bacteria suspension. Inhibition of the test sample is expressed as the minimum inhibitory concentration.

3.8. Hemagglutination Assay

1. Centrifuge human concentrated erythrocytes at 1000*g* for 3 min and wash three times with 20 m*M* Tris-HCl, 0.15*M* NaCl, pH 7.5. If a lectin tested requires calcium for hemagglutination, add an appropriate concentration of $CaCl_2$ in the same buffer.
2. Prepare 2% erythrocytes suspension (vol/vol) in the same buffer (*see* **Note 7**).
3. Mix 25 μL of the erythrocytes suspension with 25 μL of a twofold serial dilution of the lectin in a U-bottomed microtiter plate and incubate for 1 h at 37°C.
4. The titer is defined as a reciprocal value of the endpoint dilution causing hemagglutination.
5. For screening of inhibitors, premix 12.5 μL of a test sample and 12.5 μL of the lectin before adding the erythrocytes suspension. Inhibition of the sample is expressed as the minimum inhibitory concentration.

4. Notes

1. Bacteria can be stored for many years in media containing 15% glycerol at –70°C without significant loss of viability.
2. When the precipitate of bacteria is suspended and diluted with the phosphate buffer, vortex thoroughly to ensure that the bacteria are mixed and that no clumps of bacteria remain.
3. Dilute the bacteria cells by using a sterile micropipet tip to transfer 10 µL to a fresh tube containing 1 or 10 mL of the phosphate buffer. Repeat this step serially two or three times to obtain the 10^5-fold diluted suspension (5×10^3 to 1×10^4 cells/mL). As a rough guide, 1 $OD_{600} = 8 \times 10^8$ cells/mL.
4. Use at least five plates for one test sample and calculate the average value. In this assay, 300–1000 colonies can be counted on one plate in the absence of antimicrobial substances.
5. Guanofracin-Sabouraud agar for *C. albicans* is commercially available.
6. Freshly prepared sheep defibrinated blood is commercially available and can be stored for 2 wk at 4°C.
7. Volume of erythrocytes can be measured in a centrifuge tube with scale.
8. If the bacterial agglutinating activity is low, decrease the salt concentration of the buffer from 0.15–0.05*M* NaCl.
9. Out-dated human concentrated erythrocytes can be stored for at least 1 mo at 4°C unless it is diluted to 2% suspension.

Acknowledgments

This work was supported by Grants-in-Aid for Scientific Research from the Ministry of Education, Science, Sports, and Culture of Japan.

References

1. Toh, Y., Mizutani, A., Tokunaga, F., Muta, T., and Iwanaga, S. (1991) Morphology of the granular hemocytes of the Japanese horseshoe crab *Tachypleus tridentatus* and immunocytochemical localization of clotting factors and antimicrobial substances. *Cell Tissue Res.* **266**, 137–147.
2. Iwanaga, S., Muta, T., Shigenaga, T., Seki, N., Kawano, K., Katsu, T., and Kawabata, S. (1994) Structure-function relationships of tachyplesins and their analogues, in *Ciba Foundation Symposium 186, Antimicrobial Peptides* (Boman, H. G., Marsh, J., and Goode, J. A., eds.), Wiley, Chichester, England, pp. 160–175.
3. Iwanaga, S., Muta, T., Shigenaga, T., Miura, Y., Seki, N., Saito, T., and Kawabata, S. (1994) Role of hemocyte-derived granular components in invertebrate defense. *Ann. NY Acad. Sci.* **712,** 102–116.
4. Muta, T. and Iwanaga, S. (1996) Invertebrate immunology, in *Progress in Molecular and Subcellular Biology 15* (Rinkevich, B. and Müller, W. E. G., eds.), Springer-Verlag, Berlin, pp. 154–189.

5. Kawabata, S., Muta, T., and Iwanaga, S. (1996) Clotting cascade and defense molecules found in hemolymph of horseshoe crab, in *New Directions in Invertebrate Immunology* (Söderhäll, K., Iwanaga, S., and Vasta, G. R., eds.), SOS Publications, Fair Haven, NJ, pp. 255–284.
6. Saito, T., Kawabata, S., Shigenaga, T., Cho, J., Nakajima, H., Hirata, M., and Iwanaga, S. (1995) A novel big defensin identified in horseshoe crab hemocytes: isolation, amino acid sequence and antibacterial activity. *J. Biochem.* (Tokyo), **117,** 1131–1137.
7. Lehrer, R. I., Ganz, T., and Selsted, M. E. (1991) Defensins: endogenous antibiotic peptides of animal cells. *Cell* **64,** 229, 230.
8. Selsted, M. E. and Harwig, S. S. L. (1989) Determination of the disulfide array in the human defensin HNP-2. *J. Biol. Chem.* **264,** 4003–4007.
9. Selsted, M. E., Tang, Y.-Q., Morris, W. L., McGuire, P. A., Novotny, M. J., Smith, W., Henschen, A. H., and Cullor, J. S. (1993) Purification, primary structure, and antibacterial activities of ß-defensins, a new family of antimicrobial peptides from bovine neutrophils. *J. Biol. Chem.* **268,** 6641–6648.
10. Tang, Y.-Q. and Selsted, M. E. (1993) Characterization of the disulfide motif in BNBD-12, an antimicrobial ß-defensin peptide from bovine neutrophils. *J. Biol. Chem.* **268,** 6649–6653.
11. Lambert, J. Keppi, E., Dimarcq, J-L., Wicker, C., Reichhart, J-M., Dunbar, B., Lepage, P., Dorsselaer, A.V., Hoffmann, J., Fothergill, J., and Hoffmann, D. (1989) Insect immunity: isolation from immune blood of the dipteran *Phormia terranovae* of two insect antibacterial peptides with sequence homology to rabbit lung macrophage bactericidal peptides. *Proc. Natl. Acad. Sci. USA* **86,** 262–266.
12. Kuzuhara, T., Nakajima, Y., Matsuyama, K., and Natori, S. (1990) Determination of the disulfide array in sapecin, an antibacterial peptide of *Sarcophaga peregrina* (flesh fly). *J. Biochem.* (Tokyo) **107,** 514–518.
13. Saito, T., Kawabata, S., Hirata, M., and Iwanaga, S. (1995) A novel type of limulus lectin-L6: Purification, primary structure, and antibacterial activity. *J. Biol. Chem.* **270,** 14,493–14,499.
14. Okino, N., Kawabata, S., Saito, T., Hirata, M., Takagi, T., and Iwanaga, S. (1995) Purification, characterization, and cDNA cloning of a 27-kDa lectin (L10) from horseshoe crab hemocytes. *J. Biol. Chem.* **270,** 31,008–31,015.
15. Nowak, T. P. and Barondes, S. H. (1975) Agglutinin from *Limulus polyphemus*: purification with formalinized horse erythrocytes as the affinity adsorbent. *Biochim. Biophys. Acta* **393,** 115–123.
16. Bishayee, S. and Dorai, D. T. (1980) Isolation and characterization of a sialic acid-binding lectin (carcinoscorpin) from Indian horseshoe crab *Carcinoscorpius rotundicauda. Biochim. Biophys. Acta* **623**, 89–97.
17. Tsuboi, I., Matsukawa, M., Sato, N., and Kimura, S. (1993) Isolation and characterization of a sialic acid-specific binding lectin from the hemolymph of Asian horseshoe crab, *Tachypleus tridentatus. Biochim. Biophys. Acta* **1156,** 255–262.

5

Tachyplesin and Anti-Lipopolysaccharide Factor

Tatsushi Muta and Sadaaki Iwanaga

1. Introduction

Hemolymph of invertebrate animals contains many molecules involved in unique and effective innate defense systems against invading microbes. Their defense systems are activated by the recognition of common epitopes on pathogens, such as bacterial lipopolysaccharide (LPS), peptidoglycan, and fungal ß-(1,3)-glucans. In the horseshoe crab (also called limulus), an arthropod, the hemocytes are extremely sensitive to LPS on the outer membrane of Gram-negative bacteria *(1,2)*. In the presence of minute amounts of LPS, the hemocytes release coagulation factors, antimicrobial substances, and lectins by rapid exocytosis *(3–6)*. The coagulation factors constitute two types of the coagulation cascade, each of which is sensitive to LPS or ß-(1,3)-glucans *(6–8)*. Thus, after the exocytosis induced by LPS, the coagulation factors are activated on the surface of the pathogens and the pathogens are immobilized in an insoluble hemolymph clot. The pathogens are then effectively agglutinated by various types of lectin/agglutinins and finally killed by antimicrobial substances *(6,9,10)*.

Tachyplesin and anti-LPS factor (ALF) are antimicrobial peptides/proteins found from the horseshoe crab hemocytes. Both of these were originally isolated as substances that inhibited LPS-mediated activation of the LPS-sensitive coagulation factor (factor C) *(11–13)*. This inhibitory action resulted from their LPS-binding/neutralizing ability. These substances exhibit strong antibacterial activities on Gram-negative bacteria through the interaction with LPS. Tachyplesin has antimicrobial activities on Gram-negative and -positive bacteria and fungi, whereas ALF is specific to Gram-negative bacteria.

Tachyplesin is a 17-amino acid peptide with a carboxyl-terminal amide and two intramolecular disulfide bonds *(13)*. ALF contains 102 amino acid residues with an

From: *Methods in Molecular Biology, Vol. 78: Antibacterial Peptide Protocols*
Edited by: W. M. Shafer, Humana Press Inc., Totowa, NJ

intramolecular disulfide bond *(14)*. Both of these are highly basic and are stable against heat-treatment and exposure to low pH. Not only has the primary structures of these proteins/peptides from different species of horseshoe crabs been determined *(15–17)*, but their tertiary structures have also been solved *(18,19)*.

Both tachyplesin and ALF are purified from the horseshoe crab hemocytes through relatively simple procedures. Their LPS binding/neutralizing activity can be followed as the inhibitory activity on the LPS-mediated activation of factor C, which is also isolated from horseshoe crab hemocytes *(20,21)*. Although this assay is convenient, it may not be necessary since the purification procedure is highly reproducible. Because of their strong LPS-binding/neutralizing activity, they would be useful in the field of basic science as well as in clinical studies; perhaps they might prove useful in the treatment of Gram-negative sepsis.

2. Materials (*see* Note 1)

2.1. Preparation of Horseshoe Crab Hemocyte Lysate and Debris

1. 70% (v/v) ethanol.
2. Pyrogen-free distilled water (Otsuka Pharmaceutical, Tokyo).
3. 20 m*M* Tris-HCl, pH 8.0, 3% NaCl, 100 m*M* caffeine.
4. 20 m*M* Tris-HCl, pH 8.0, 3% NaCl.
5. 20 m*M* Tris-HCl, pH 8.0, 50 m*M* NaCl.

2.2. Purification of Anti-LPS factor

1. Sepharose CL-6B (Pharmacia, Uppsala, Sweden).
2. Dextran sulfate (Pharmacia).
3. BrCN.
4. Acetonitrile.
5. 10*M* NaOH.
6. 20 m*M* Tris-HCl, pH 8.0.
7. 20 m*M* Tris-HCl, pH 8.0, 50 m*M* NaCl.
8. 20 m*M* Tris-HCl, pH 8.0, 0.3*M* NaCl.
9. 20 m*M* Tris-HCl, pH 8.0, 0.5*M* NaCl.
10. 20 m*M* Tris-HCl, pH 8.0, 2.0*M* NaCl.
11. Sephadex G-50 (Pharmacia).
12. 20 m*M* HCl.
13. CM-Sepharose CL-6B (Pharmacia).
14. 50 m*M* sodium acetate, pH 4.9.
15. 50 m*M* sodium acetate, pH 4.9, 1.5*M* NaCl.

2.3. Purification of Tachyplesin

1. 20 m*M* HCl.
2. Sephadex G-50 (Pharmacia).
3. 20 m*M* HCl.

4. CM Sepharose CL-6B (Pharmacia).
5. 1*M* NaOH.
6. 20 m*M* sodium acetate, pH 6.0.
7. 20 m*M* sodium acetate, pH 6.0, 1.5*M* NaCl.

2.4. Anti-LPS Factor or Tachyplesin Assay (LPS-Neutralizing Activity)

1. 50 m*M* Tris-HCl, pH 8.0.
2. 1 μg/mL LPS from *Escherichia coli* 0111:B4 (from List Biological Laboratories, Campbell, CA) (*see* **Note 13**).
3. Horseshoe crab factor C. Purified as described ***(20)***.
4. 10 mg/mL bovine serum albumin (BSA) (essentially fatty acid-free BSA, from Sigma, St. Louis, MO).
5. 1*M* Tris-HCl, pH 8.0.
6. Factor C mixture. Mix 1 μL of 0.6–1.2 mg/mL factor C, 5 μL of 10 mg/mL BSA, and 15 μL of 1*M* Tris-HCl, pH 8.0, and 39 μL of pyrogen-free distilled water. Prepare fresh mixture just before use and keep on ice.
7. 2 m*M* Boc-Val-Pro-Arg-pNA (the Protein Research Foundation, Osaka) in pyrogen-free distilled water.
8. 0.6*M* acetic acid.

2.5. Antibacterial Activity Assay

1. JY media (synthetic Jarvis's media), 7.5 g/L glucose, 0.02 g/L sodium-L-glutamate, 0.1 mg/L calcium pantothenate, 1.25 g/L $(NH_4)_2SO_4$, 3.0 g/L KH_2PO_4, 6.0 g/L Na_2HPO_4, 0.30 g/L $MgSO_4 \cdot 7H_2O$, 5 g/L NaCl ***(22)*** supplemented with 2 mg/mL of yeast extract (Difco, E. Molesley, Surrey, UK).

3. Methods

3.1. Preparation of Horseshoe Crab Hemocyte Lysate and Debris

3.1.1. Bleeding of the Horseshoe Crab Hemolymph by Cardiac Puncture

1. Rinse the joint between the cephalothorax and the abdomen with 70% ethanol and, then, with sterilized distilled water.
2. Bleed the hemolymph by inserting a sterilized needle into the joint.
3. Collect the hemolymph into a sterilized container containing one-tenth volume of cold 20 m*M* Tris-HCl, pH 8.0, containing 3% NaCl and 100 m*M* caffeine. Quickly chill on ice. Pool 200–300 mL of clean hemolymph into a sterilized centrifugation tube (*see* **Note 2**).

3.1.2. Preparation of Hemocytes from the Hemolymph

1. Centrifuge at 3000*g* for 15 min at 4°C. Discard the supernatant.
2. Resuspend the pellet in 250 mL of 20 m*M* Tris-HCl, pH 8.0, containing 3% NaCl and centrifuge at 3000*g* for 30 min at 4°C. Discard the supernatant.

3. Repeat **step 2**.
4. Measure the weight of the hemocyte pellet and store them at –80°C until homogenized to prepare lysate. It is stable at least for 2 yr.

3.1.3. Preparation of the Hemocyte Lysate and Debris

1. Thaw frozen hemocytes (50–100 g) in running water. Once thawed, immediately stand on ice.
2. Suspend the hemocytes in 200 mL of 20 m*M* Tris-HCl, pH 8.0, containing 50 m*M* NaCl in a sterilized centrifugation tube.
3. Homogenize the suspension with Physcotron® for 3 min (*see* **Notes 3–5**).
4. Centrifuge at 12,000*g* at 4°C for 30 min. Transfer the supernatant into a sterilized container.
5. Re-extract the pellet three times more with 200 mL of the same buffer. Collect the supernatant and pool them (lysate) (*see* **Note 6**).
6. Freeze pellet as hemocyte debris at –80°C unless it is used in the same day. It is stable at least for 2 yr.

3.2. Purification of Anti-LPS Factor

3.2.1. Preparation of Dextran Sulfate-Sepharose CL-6B

1. Wash Sepharose CL-6B (250 mL) with 2 L of distilled water. Transfer to a beaker.
2. Add 500 mL of ice-cold 50 mg/mL dextran sulfate. Mix well.
3. Add 120 mL of 0.5 g/mL BrCN in acetonitrile.
4. Incubate the mixture on ice for 45 min with keeping the pH at 10.5 and the temperature at 4–10°C by dropping 10*M* NaOH and ice made from sterilized distilled water.
5. Wash the resin with more than 2 L of sterilized distilled water.
6. Suspend the resin in 500 mL of 20 m*M* Tris-HCl, pH 8.0, and incubate at 4°C for more than 12 h.

3.2.2. Dextran Sulfate-Sepharose CL-6B Column Chromatography

1. Equilibrate a dextran sulfate-Sepharose CL-6B column (approx 300 mL) with 20 m*M* Tris-HCl (pH 8.0) containing 50 m*M* NaCl.
2. Apply hemocyte lysate (600–800 mL) onto a column. Start to collect the eluate.
3. Wash the column with three column volumes of the equilibration buffer.
4. Wash the column with 20 m*M* Tris-HCl, pH 8.0, containing 0.3, 0.5, and 2.0*M* NaCl in a stepwise fashion. Anti-LPS factor is eluted in the last 2*M* NaCl fraction (*see* **Note 7**).
5. Pool the 2*M* NaCl fraction and lyophilize.

3.2.3. Sephadex G-50 Column Chromatography (see **Note 8**)

1. Dissolve the lyophilized 2*M* NaCl fraction of dextran sulfate-Sepharose CL-6B column chromatography in a minimum volume of 20 m*M* HCl.

2. Centrifuge at 12,000*g* for 10 min at 4°C to remove insoluble materials. Transfer the supernatant into a new tube.
3. Apply the supernatant onto a Sephadex G-50 column (800–1000 mL) pre-equilibrated with 20 m*M* HCl.
4. Elute proteins with 20 m*M* HCl. Collect fractions.
5. Pool fractions exhibiting ALF activity.

*3.2.4. CM-Sepharose CL-6B Column Chromatography (Optional) (see **Notes 9–11**)*

1. Lyophilize ALF-containing fraction.
2. Dissolve the lyophilized material in 50 m*M* sodium acetate, pH 4.9.
3. Load the sample onto a CM-Sepharose CL-6B (30–50 mL) pre-equilibrated with 50 m*M* sodium acetate, pH 4.9.
4. Wash the column with the equilibration buffer.
5. Elute bound proteins with a linear gradient of NaCl from 0 to 1.5*M*.
6. Pool the fractions containing pure ALF.

3.3. Purification of Tachyplesin

3.3.1. Acid Extraction of Debris

1. Thaw frozen hemocyte debris (approx 50 g) in running water. Once thawed, immediately stand on ice.
2. Suspend the hemocyte debris in 100 mL of 20 m*M* HCl in a sterilized tube.
3. Homogenize the suspension with Physcotron® for 3 min.
4. Centrifuge at 12,000*g* at 4°C for 30 min. Transfer the supernatant into a sterilized container.
5. Re-extract the pellet twice more with 100 mL of 20 m*M* HCl. Collect the supernatant and pool them (acid extract).
6. Lyophilize the acid extract to concentrate.

*3.3.2. Sephadex G-50 Column Chromatography (see **Note 12**)*

1. Dissolve the lyophilized acid extract in 50 mL of 20 m*M* HCl.
2. Apply the sample to a Sephadex G-50 column (700–1000 mL) pre-equilibrated with 20 m*M* HCl and elute proteins with 20 m*M* HCl.
3. Pool fractions containing tachyplesin activity.

3.3.3. CM Sepharose CL-6B Column Chromatography

1. Adjust pH of the sample to 6.0 with 1*M* NaOH.
2. Apply the sample to a CM-Sepharose CL-6B (~50 mL) pre-equilibrated with 20 m*M* sodium acetate, pH 6.0.
3. Wash the column extensively with the same buffer.
4. Elute tachyplesin with a linear gradient from 500 mL each of the same buffer with and without 1.5*M* NaCl.
5. Pool the fraction containing pure tachyplesin.

3.4. Anti-LPS Factor or Tachyplesin Assay (LPS-Neutralizing Activity) (see Notes 14 and 15)

1. Dilute sample more than 10 times with 50 m*M* Tris-HCl (pH 8.0).
2. Incubate 100 µL of diluted sample with 40 µL of 1 µg/mL LPS at 37°C for 5 min.
3. Add 60 µL of 10–20 µg/mL factor C mixture and incubate at 37°C for 15 min.
4. Add 50 µL of 2 m*M* Boc-Val-Pro-Arg-pNA and incubate for 5–10 min.
5. Terminate the reaction by adding 800 µL of 0.6*M* acetic acid.
6. Measure absorbance at 405 nm of the resulting solution.

3.5. Antibacterial Activity Assay (see Note 16)

1. Add 20 µL of serial dilutions of sample to mixtures of 20 µL of bacterial suspension (approx 10^6/mL) and 160 µL of JY media in each well of a flat-bottomed microplate.
2. Incubate at 37°C for 18–20 h.
3. Read absorbance at 550 nm on a microplate reader taking the absorbance of a well filled with uninoculated media as a reference.

4. Notes

1. All glassware and metalware used in the purification and assay should be sterilized by heating at 220°C for 3 h or by soaking overnight in 95% ethanol containing 0.2*M* NaOH. All the buffer solutions are made up with pyrogen-free distilled water and sterilized by autoclaving for more than 30 min. The buffers are chilled at 4°C before use.
2. The bled hemolymph (*see* **Subheading 3.1.1.**) should be left on ice for more than 5 min to confirm that it does not form a clot caused by a trace amount of contamination. The contaminated hemolymph should be discarded. Never pool with clean hemolymph.
3. We use Physcotron® purchased from Niti-on Medical and Physical Instruments, Chiba, Japan. Other homogenizer such as Polytron® would work as well.
4. The blade of a Physcotron is sterilized overnight by soaking in 95% ethanol containing 0.2*M* NaOH.
5. The homogenization with Physcotron for longer time may heat up the lysate. The lysate should be chilled on ice before centrifugation.
6. When large amounts of hemocytes are used, some aggregated proteins in the lysate could be stuck on the column. In order to precipitate those aggregates, the lysate may stand overnight at 4°C and then be centrifuged again just before loading to a dextran sulfate-Sepharose CL-6B column.
7. On dextran sulfate Sepharose CL-6B column chromatography, the 2*M* NaCl fraction (*see* **Subheading 3.2.2.**) may split into two protein peaks. The second peak should contain anti-LPS factor. We experienced this phenomenon when American horseshoe crabs are used.
8. On Sephadex-G50 column chromatography, the second peak of absorbance at 280 nm should contain anti-LPS factor (*see* **Subheading 3.2.3.**). This prepa-

ration usually has more than 80% homogeneity. However, if more purity is desired, further purification can be performed by a CM-Sepharose CL-6B column.

9. After the CM-Sepharose CL-6B column chromatography (*see* **Subheading 3.2.4.**), the purified sample can be desalted by passing a small column of Sephadex G-25 in 20 m*M* HCl. Eluted anti-LPS factor can be lyophilized and dissolved in desired buffer.
10. The extinction coefficient of anti-LPS factor (*see* **Subheading 3.2.4.**) in 20 m*M* HCl was calculated to be $A_{280}^{1\%} = 27.4$. Purified anti-LPS factor is stable on heat-treatments in neutral (pH 7.5) and acidic (pH 1.0) media at 100°C for 5 min without apparent loss of the activity.
11. If the highest purity (*see* **Subheading 3.2.4.**) is required for studies such as sequence analysis, it can be further purified with a reversed-phase HPLC using a buffer system of 0.1% trifluoroacetic acid/acetonitrile on a C18 column such as Cosmosil $5C_{18}$-P (3.9 × 300 mm).
12. After Sephadex G-50 column chromatography, tachyplesin can be purified with more than 80% homogeneity. It can be further purified by CM-Sepharose CL-6B column chromatography or by a reversed-phase HPLC as anti-LPS factor.
13. LPS from *E. coli* 0111:B4 is usually dissolved in sterilized distilled water. If it is difficult to dissolve, add small amount of triethylamine. Dissolved LPS is heated at 70°C for 3–5 min. Sonication may help its dissolution.
14. The ALF and tachyplesin activities (*see* **Subheading 3.4.**) are expressed as the inhibitory activity of the LPS-mediated activation of factor C.
15. The concentration of factor C and/or incubation time after the addition of the substrate should be adjusted as absorbance at 405 nm of a control (50 m*M* Tris-HCl, pH 8.0, instead of sample) does not exceed 0.7 (*see* **Subheading 3.4.3.**). When it is more than 0.7, the reaction is not quantitative.
16. MIC (minimal inhibitory concentration) is expressed as the lowest final concentration at which no growth was observed (*see* **Subheading 3.5.**).

Acknowledgments

This work was supported by Grants-in-Aid for Scientific Research from the Ministry of Education, Science, and Culture of Japan.

References

1. Bang, F. B. (1956) A bacterial disease of *Limulus polyphemus. Bull. Johns Hopkins Hosp.* **98,** 325–350.
2. Levin, J. and Bang, F. B. (1964) The role of endotoxin in the extracellular coagulation of Limulus blood. *Bull. Johns Hopkins Hosp.* **115,** 265–274.
3. Ornberg, R. L. and Reese, T. S. (1979) Secretion in *Limulus* amebocytes is by exocytosis. *Prog. Clin. Biol. Res.* **29,** 125–130.
4. Toh, Y., Mizutani, A., Tokunaga, F., Muta, T., and Iwanaga, S. (1991) Morphology of the granular hemocytes of the Japanese horseshoe crab *Tachypleus*

tridentatus and immunocytochemical localization of clotting factors and antimicrobial substances. *Cell Tissue Res.* **266,** 137–147.

5. Iwanaga, S., Muta, T., Shigenaga, T., Miura, Y., Seki, N., Saito, T., and Kawabata, S. (1994) Role of hemocyte-derived granular components in invertebrate defense. *Ann. NY Acad. Sci.* **712,** 102–116.
6. Muta, T. and Iwanaga, S. (1996) Clotting and immune defense in limulidae, in *Progress in Molecular and Subcellular Biology*, vol. 15, *Invertebrate Immunology* (Rinkevich, B., and Müller, W. E. G., eds.) Springer-Verlag, Berlin, Germany, pp. 154–189.
7. Iwanaga, S., Miyata, T., Tokunaga, F., and Muta, T. (1992) Molecular mechanism of hemolymph clotting system in Limulus. *Thrombos. Res.* **68,** 1–32.
8. Iwanaga, S. (1993) The limulus clotting reaction. *Curr. Opin. Immunol.* **5,** 74–82.
9. Muta, T., Nakamura, T., Furunaka, H., Tokunaga, F., Miyata, T., Niwa, M., and Iwanaga, S. (1990) Primary structures and functions of anti-lipopolysaccharide factor and tachyplesin peptide found in horseshoe crab hemocytes. *Adv. Exp. Med. Biol.* **256,** 273–285.
10. Iwanaga, S., Muta, T., Shigenaga, T., Seki, N., Kawano, K., Katsu, T., and Kawabata, S. (1994) Structure-function relationships of tachyplesins and their analogues, in *Ciba Foundation Symposium*, vol. 186, *Antimicrobial Peptides.* (Boman, H. G., Marsh, J. and Goode, J. A., eds.), Wiley, Chichester, England, pp. 160–175.
11. Tanaka, S., Nakamura, T., Morita, T., and Iwanaga, S. (1982) *Limulus* anti-LPS factor: an anticoagulant which inhibits the endotoxin-mediated activation of *Limulus* coagulation system. *Biochem. Biophys. Res. Commun.* **105,** 717–723.
12. Morita, T., Ohtsubo, S., Nakamura, T., Tanaka, S., Iwanaga, S., Ohashi, K., and Niwa, M. (1985) Isolation and biological activities of Limulus anticoagulant (anti-LPS factor) which interact with lipopolysaccharide (LPS). *J. Biochem.* (Tokyo) **97,** 1611–1620.
13. Nakamura, T., Furunaka, H., Miyata, T., Tokunaga, F., Muta, T., Iwanaga, S., Niwa, M., Takao, T., and Shimonishi, Y. (1988) Tachyplesin, a class of antimicrobial peptide from the hemocytes of the horseshoe crab *(Tachypleus tridentatus)*. Isolation and chemical structure. *J. Biol. Chem.* **263,** 16,709–16,713.
14. Aketagawa, J., Miyata, T., Ohtsubo, S., Nakamura, T., Morita, T., Hayashida, H., Miyata, T., Iwanaga, S., Takao, T., and Shimonishi, Y. (1986) Primary structure of Limulus anticoagulant anti-lipopolysaccharide factor. *J. Biol. Chem.* **261,** 7357–7365.
15. Miyata, T., Tokunaga, F., Yoneya, T., Yoshikawa, K., Iwanaga, S., Niwa, M., Takao, T., and Shimonishi, Y. (1989) Antimicrobial peptides, isolated from horseshoe crab hemocytes, tachyplesin II, and polyphemusins I and II. Chemical structures and biological activity. *J. Biochem.* (Tokyo) **106,** 663–668.
16. Muta, T., Fujimoto, T., Nakajima, H., and Iwanaga, S. (1990) Tachyplesins isolated from hemocytes of Southeast Asian horseshoe crabs *(Carcinoscorpius rotundicauda* and *Tachypleus gigas)*: identification of a new tachyplesin, tachy-

plesin III, and a processing intermediate of its precursor. *J. Biochem.* (Tokyo) **108,** 261–266.
17. Muta, T., Miyata, T., Tokunaga, F., Nakamura, T., and Iwanaga, S. (1987) Primary structure of anti-lipopolysaccharide factor from American horseshoe crab, *Limulus polyphemus. J. Biochem.* (Tokyo) **101,** 1321–1330.
18. Kawano, K., Yoneya, T., Miyata, T., Yoshikawa, K., Tokunaga, F., Terada, Y., and Iwanaga, S. (1990) Antimicrobial peptide, tachyplesin I, isolated from hemocytes of the horseshoe crab *(Tachypleus tridentatus).* NMR determination of the beta-sheet structure. *J. Biol. Chem.* **265,** 15,365–15,367.
19. Hoess, A., Watson, S., Siber, G. R., and Liddington, R. (1993) Crystal structure of an endotoxin-neutralizing protein from the horseshoe crab, Limulus anti-LPS factor, at 1.5 Å resolution. *EMBO J.* **12,** 3351–3356.
20. Nakamura, T., Morita, T., and Iwanaga, S. (1986) Lipopolysaccharide-sensitive serine-protease zymogen (factor C) found in Limulus hemocytes. Isolation and characterization. *Eur. J. Biochem.* **154,** 511–521.
21. Muta, T., Tokunaga, F., Nakamura, T., Morita, T., and Iwanaga, S. (1993) Limulus clotting factor C: lipopolysaccharide-sensitive serine protease zymogen. *Methods Enzymol.* **223,** 336–345.
22. Jarvis, F. G., Mesenko, M. T., and Kyte, J. E. (1960) Electrophoretic purification of the Vi antigen. *J. Bacteriol.* **80,** 677–682.

6

Circular Dichroism Studies of Secondary Structure of Peptides

Martha M. Juban, Maryam M. Javadpour, and Mary D. Barkley

1. Introduction

The optical activity characteristic of organic molecules is a result of the absorption of light as electrons are promoted to higher molecular orbitals. In UV-visible spectroscopy, the transition process is described by the Beer-Lambert Law, A = εcd, in which A is the experimentally measured absorbance, ε is the extinction coefficient, c is the molar concentration of solute, and d is the path length. Absorbance, A, is plotted vs energy or wavelength of the light.

Plane-polarized light can be resolved into its two circularly polarized components: left circularly polarized light, whose electric vector rotates counterclockwise about the axis perpendicular to the direction of travel of the light beam, and right circularly polarized light, whose rotation is clockwise. A chiral compound exhibits optical activity because its absorption of left circularly polarized light is not equal to its absorption of right circularly polarized light. After passing through a chiral medium, the electric vectors describe an ellipse whose major axis lies along a new angle of rotation. The measured eccentricity of the ellipse represents the unequal absorptions of left and right circularly polarized light referred to as circular dichroism, CD. Written as analagous to Beer's Law, the experimentally measured ellipticity = θcd, in which θ is the molar value in radians. A CD spectrum plots the differential absorbance of left and right circularly polarized light vs wavelength.

Circular dichroism is useful in determining the secondary structure of peptides because we can reasonably assume that in the absence of aromatic amino acid residues only the peptide backbone contributes significantly in the far UV

From: *Methods in Molecular Biology, Vol. 78: Antibacterial Peptide Protocols*
Edited by: W. M. Shafer, Humana Press Inc., Totowa, NJ

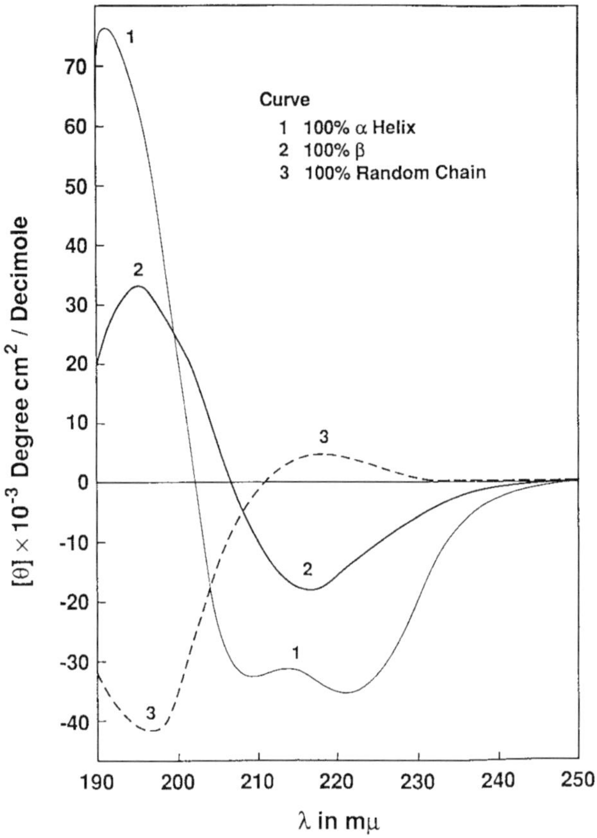

Fig. 1. Circular dichroism spectra of poly-L-lysine in the α-helical, β, and random conformations. Reprinted with permission from **ref. *5***. Copyright (1969) American Chemical Society.

region and because the spectrum reflects the spatial arrangement of chiral units in the peptide chain. The presence of aromatic amino acids induces a positive CD signal, causing significant errors in ellipticity values ***(1,2)***. There are three main classes of secondary structure: the alpha helix, the beta sheet, and the random coil. The α-helix produces the most distinctive CD spectrum: a very strong positive band near 192 nm, which corresponds to $\pi \rightarrow \pi^*_\perp$ transitions, and two negative maxima of approximately equal intensity near 222 nm ($n \rightarrow \pi^*$) and 208 nm ($\pi \rightarrow \pi^*_{\parallel}$). The beta sheet exhibits a single negative band near 217 nm, representing $n \rightarrow \pi^*$. The random coil exhibits a strong negative band at 197 nm ($\pi \rightarrow \pi^*_{\parallel}$) and a small positive band at 217 nm ($n \rightarrow \pi^*$) (*see* **Fig. 1**).

Peptide conformation is highly sensitive to solution variables; a peptide may be insoluble in one condition, but soluble under other conditions. In

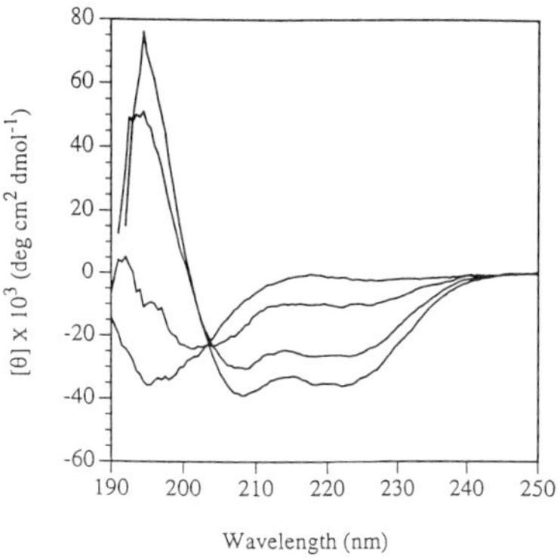

Fig. 2. As [NaCl] increases from 0–0.6*M*, the peptide conformation changes from random coil to α-helix. The isodichroic point indicates a two-state system.

characterizing the secondary structure of peptides by CD, the effects of pH, ionic strength of the buffer, and peptide concentration should be examined. As a general rule, aggregation increases with peptide concentration, ionic strength, or pH. Concentration dependence reveals whether the peptide is monomeric or self-associating: for example, a shift from random coil to α-helix as the peptide concentration increases indicates an association equilibrium. Self-association may also be obtained by changing the pH (generally increasing) to titrate the amino groups present in the peptide sequence, or by changing the ionic strength of the buffer to optimize charge screening. Variations in these parameters are examined systematically in an effort to determine the number and types of species present in the solution. Stable equilibria between only two states are identified by the presence of an isodichroic point (*see* **Fig. 2**).

The concentration dependence of possible transitions should be studied by CD prior to undertaking sedimentation equilibrium studies in the analytical ultracentrifuge. The choice of solution conditions that yield an isodichroic point will result in a well-behaved system for further study.

2. Materials

1. Instruments capable of measuring the CD spectrum down to 180 nm, such as the Aviv Model 62DS (Lakewood, NJ) or Jasco J-40 (Easton, MD), are required for reliable results *(3)*.
2. Both cylindrical and rectangular quartz cells are commercially available in a variety of path lengths from Hellma (Jamaica, NY). We use rectangular cells from Aviv that have been prescreened for transparency to circularly polarized light (*see* **Note 1**).

cell path length, cm	approximate sample volume, μL
1.0	3000
0.1	300
0.05	120
0.01	50
0.001	30

3. Buffers should be chosen based on the solubility and stability of the sample of interest, and on the transparency of the buffer at the wavelengths at which data will be collected (for peptides, 250–190 nm).
4. (+)-1O-camphorsulfonic acid (Aldrich, Milwaukee, WI) for calibration of the instrument ***(3)***.
5. Teflon syringe needle with a Luer hub (Aldrich) for cleaning and drying cells (*see* **Note 2**).

3. Method

3.1. Instrument Preparation

1. Before turning on the high-intensity lamp, it is essential to remove all oxygen from the instrument that might be converted into ozone and damage the optical system. Purge with nitrogen for at least 15–30 min before turning on the lamp.
2. The instrument should be calibrated periodically with an aqueous solution of camphorsulfonic acid, 1 mg/mL in a 1-mm cell ***(3)***.

3.2. Sample Preparation

1. Purified, dry peptide is weighed, then dissolved in the buffer of choice. Accurate concentrations are very important, and calculations should not rely simply on the apparent dry mass. Quantitative amino acid analysis to determine the percent of peptide mass is required for peptides that do not contain tryptophan or tyrosine. Concentrations of peptides containing tryptophan and tyrosine may be accurately measured by absorbance at 280 nm if the sequence is accurately known, because the only residues that contribute significantly to the measured optical density at 280 nm are tryptophan, tyrosine, and cystine. The extinction coefficient may be calculated as:

$$\varepsilon_{280} = a\ \varepsilon_{tyr} + b\ \varepsilon_{trp} + c\ \varepsilon_{cys}$$

where a, b, and c are the number of each type of residue per molecule, ε_{tyr} is $1280M^{-1}\ cm^{-1}$, ε_{trp} is $5690M^{-1}\ cm^{-1}$, and ε_{cys} is $120M^{-1}\ cm^{-1}$ ***(4)***.

2. For ionic strength or pH studies, the concentration of the peptide should be constant while the salt concentration or pH of the buffer is varied.
3. For concentration studies, the buffer conditions should remain constant throughout the experiment while the peptide concentration is varied.
4. Prepared samples should be checked on a spectrophotometer for an absorbance between 0.1 and 1.0 at the wavelength of maximum absorption before taking the

time to measure CD. If the absorbance is too low, try a cell with a longer path length. If the absorbance is too high, dilute the sample or use a shorter path length cell. An absorbance of 0.7 is usually optimal.

3.3. Data Collection

1. Scan the sample from 250 down to 180 nm, or as far down as the increase in photomultiplier (or dynode) voltage will allow. Typical parameters would be a scan from 250–190 nm, using a step size of 0.5 nm, a time of 5 s at each wavelength, and averaging three scans. For an acceptable signal-to-noise ratio, the voltage should remain under 600 V. If the voltage increases, increase the band width to 0.7 or 1.0 nm.
2. Obtain a baseline scan using the same parameters on buffer alone.

3.4. Data analysis

1. Subtract the baseline scan of buffer alone from the data scan of the sample.
2. The corrected data are expressed in millidegrees and should be converted to molar ellipticity using the formula $[\theta] = [\theta]_{obs}(MRW/10cd)$, where $[\theta]_{obs}$ is the experimentally measured ellipticity, MRW is the mean residue molecular weight of the peptide (molecular weight divided by the number of peptide bonds), *c* is the concentration of the sample in mg/mL, and *d* is the optical pathlength of the cell in cm. If the peptide is a C-terminal amide or if the N-terminus is acetylated, count the terminus as an additional residue.
3. Three accepted methods exist for calculating the percent helicity in a two-state system.
   a. Percent α-helix = $[(\Theta)_{208\ nm} - 4000]/[33{,}000 - 4{,}000]$ ***(5)***.
   b. Determine experimental fractional helix contents from $[\Theta]_{222}$ measurements using –40,000 (1–2.5*n*) and 0 deg cm^2/dmol as the values for 100 and 0% helix, respectively, *n* is the number of amino acid residues in the peptide ***(6)***.
   c. Percent α-helix = $-100\ (\Theta_{222} + 3000)/33000$ ***(7)***.
4. The percentages of the typical secondary elements making up the peptide structures (α-helix, β-pleated sheet, and so on) are determined by Gerald D. Fasman's LINCOMB-CCA CD analysis program. LINCOMB is a simple algorithm based on a least-square fit with a set of reference spectra representing the known secondary structures and yielding an estimation of weights attributed to α-helix, β-sheet, and random coil contributions of the peptide being analyzed. The Convex Constraint Algorithm (CCA) CD analysis program is a general deconvolution method for a CD spectrum of any combination of secondary structures. The CCA program deconvolutes a set of CD spectra data stored in the "A[MxN]" matrix. The procedure results in a given number of "pure" (or component) spectra stored in the "B[MxS]" matrix and simultaneously creates the adjacent coefficient matrix (called conformational weight "C[SxN]" matrix), which determines the contribution of these "pure" spectra to the CD spectrum of the peptide.

4. Notes

1. Cells smaller than 1-mm path length are generally demountable, consisting of two glass plates with a layer of frosting that holds them the correct distance apart. If such cells do not seal well, use a small strip of Parafilm around the perimeter of the plates.
2. Cells should be cleaned with detergent and rinsed a minimum of 10 times with distilled water after each sample. They should be thoroughly dried before adding the next sample to prevent errors in calculating concentration. Cleaning and drying cells is facilitated by a Teflon syringe needle with a luer hub connected through a 1-mL syringe to a water aspirator.

References

1. Manning, M. C. and Woody, R. W. (1989) Theoretical study of the contribution of aromatic side chains to the circular dichroism of basic bovine pancreatic trypsin inhibitor. *Biochemistry* **28,** 8609–8613.
2. Chakrabartty, A., Kortemme, T., Padmanabhan, S., and Baldwin, R. (1993) Aromatic side chain contribution to far-ultraviolet circular dichroism of helical peptides and its effects on measurement of helix propensities. *Biochemistry* **32,** 5560–5565.
3. Johnson, W. C. Jr. (1990) Protein secondary structure and circular dichroism: a practical guide. *Proteins: Struc. Func. Genet.* **7,** *205–214.*
4. Gill, S. C. and von Hippel, P. H. (1989) Calculation of protein extinction coefficients from amino acid sequence data. *Analyt. Biochem.* **182,** 319.
5. Greenfield, N. and Fasman, G. D. (1969) Computed circular dichroism spectra for the evaluation of protein conformation. *Biochemistry* **8,** 4108–4116.
6. Chakrabartty, A., Schellman, J. A., and Baldwin, R. L. (1991) Large differences in the helix propensities of alanine and glycine. *Nature* **351,** 586–588.
7. McLean, L. R., Hagaman, K. A., Owen, T. J., and Krstenansky, J. L. (1991) Minimal peptide length for interaction of amphipathic a-helical peptides with phosphatidylcholine liposomes. *Biochemistry* **30,** 31–37.
8. Perczel, A., Park, K., and Fasman, G. D. (1992) Analysis of the circular dichroism spectrum of proteins using the convex constraint algorithm: a practical guide. *Analyt. Biochem.* **203,** 83–93.

7

Analytical Ultracentrifugation Studies of Association of Peptides

Martha M. Juban, Maryam M. Javadpour, and Mary D. Barkley

1. Introduction

Sedimentation equilibrium experiments in the Beckman Optima XL-A Analytical Ultracentrifuge (Fullerton, CA) can be used to determine aggregation states of synthetic peptides in homogeneous associating systems such as $nP \overset{K}{\leftrightarrow} P_n$ in which P is peptide monomer and $K = [P_n]/[P]^n$. Experimental conditions are chosen so that transport by sedimentation is balanced by diffusional transport within a sample. The equilibrium concentration distribution formed over a period of several hours is analyzed using known monomeric molecular weights to determine the aggregation number and thermodynamic parameters such as equilibrium constants, ΔG, ΔH, ΔS, and ΔCp. It is preferable to use three different concentrations of the same peptide, run simultaneously, and to achieve equilibria at several different temperatures.

Data analysis should begin with the simplest system possible (a single, ideal component) and progress to more complex reversible equilibria of the type monomer to "n"-mer until a satisfactory fit is achieved. For an ideal single component, the equilibrium concentration distribution is an exponential function described by **Eq. 1:**

$$C_r = C_b \exp [AM(r^2 - r_b^2)] + \varepsilon \qquad (1)$$

in which C_r is the concentration at radial position r in the cell; C_b is a reference concentration of monomer at the arbitrarily selected radius of the cell bottom, designated r_b; $A = (1 - \underline{v}\rho)\omega^2/2RT$, where $\underline{v}$ is the partial specific volume of the macromolecule, ρ is the solution density, ω is the rotor angular velocity, R is

From: *Methods in Molecular Biology, Vol. 78: Antibacterial Peptide Protocols*
Edited by: W. M. Shafer, Humana Press Inc., Totowa, NJ

the gas constant, and T is the absolute temperature; M is the molecular mass of the monomer; and ε is a small baseline error term.

For a system in which two interacting species are present, monomer and "n"-mer, the mathematical model describing the association will be **Eq. 2:**

$$c_{r,i} = c_b \exp [AM(r^2-r_b^2)] + c_b^n \exp [\ln k_{1,n} + nAM(r^2-r_b^2)] + \varepsilon \qquad (2)$$

where ln k is the natural logarithm of the association constant k written on an absorbance scale. When the model being tested includes more than one associated state, the equation becomes more complex, the most general form being **Eq. 3:**

$$c_r = c_b \exp [AM(r^2-r_b^2)] + \sum_{2\leq i\leq n} c_b^i \exp [\ln k_{1,i} + iAM(r^2-r_b^2)] + \varepsilon \qquad (3)$$

2. Materials

2.1. Analytical cells

The An-60Ti rotor holds three sample cells in addition to a counterbalance. The cell consists of a two- or six-channel centerpiece stacked between two quartz window assemblies inside a cylindrical housing. Epon centerpieces are normally used for aqueous solutions of pH 3–10; aluminum centerpieces are required for organic solvents or extreme pH. Epon two-channel centerpieces have a fill volume of 0.45 mL/channel, and give a column length of up to 14 mm. This longer column yields more data points but requires more time to come to equilibrium, since the equilibration time is proportional to the square of the column length. The six-channel centerpiece has a fill volume of 0.12 mL/channel, giving a liquid column about 3 mm high or less from base to meniscus. The six-channel design allows the centrifugation of three sample-solvent pairs simultaneously in one cell, and the short column height greatly reduces the time required to reach equilibrium (*see* **Note 1**).

2.2. Buffers

Buffers should be chosen based on the solubility and stability of the sample of interest, and on the transparency of the buffer at the wavelengths at which data will be collected.

2.3. Computer Requirements

Beckman recommends an IBM-compatible computer with a 486 processor and at least 8 MB RAM to run the XL-A program. The data-analysis software should be capable of performing nonlinear least-squares curve-fitting procedures and have good graphics capabilities.

3. Method

3.1. Preparation of Sample

1. Purified, dry peptide is weighed, then dissolved in the buffer of choice. Calculations of the concentration should take into account the percent peptide mass of the lyophilized powder and not rely simply on the apparent dry mass.
2. The peptide solution used in each cell should have an absorbance of 0.2–0.4 at the selected wavelength. It is preferable to check sample absorbance on a spectrophotometer before loading the cells.
3. Reserve a corresponding volume of solvent to use as a reference solution (*see* **Note 2**).

3.2. Preliminary Calculations

1. Use of **Eqs. 1–3** requires knowledge of the solvent density (in most instances, the density of a dilute solution is approximated by using the solvent density) and the partial specific volume of the solute, which may be either calculated or measured directly. Estimates of these values are necessary to determine an appropriate rotor speed before beginning the equilibrium experiment. Solvent density must be corrected to the experimental temperature. If a density meter like the PAAR DMA 58 is available, direct measurement is preferable. If solvent density must be calculated, begin with density values for pure water *(1)* and estimate by summing the density increments for each component as listed in Table 3, p. 106 (in **ref.** *2*).
2. Partial specific volume is defined as the change in volume of the solution per gram of solute. A useful average value for all proteins is 0.72 mL/g. Partial specific volume may be calculated from the sequence with an accuracy of 1–2%, using values from Table 2a, p. 98 (in **ref.** *2*). It may be measured directly by doing parallel experiments in deuterium oxide and water *(3)*.
3. An optimum rotor speed will result in a concentration gradient 4–10 times greater at the bottom of the cell than at the top when equilibrium is achieved. An estimate may be determined from the nomograph published by Beckman *(4)*. It is helpful to enter the data analysis program at this point, using estimated values for all the parameters in the appropriate equation, and draw the plots that represent the expected aggregation states at equilibria achieved at various speeds.

3.3. Loading Cells

1. The cells are assembled of scrupulously clean components and the screw-ring at the top torqued to 120 in.-pounds using the Beckman collett torque wrench.
2. With the screw-ring end facing you, and the fill holes up, the sector on the right is filled with 100–300 μL of sample. Adjustable pipets with disposable gel-loading tips work well if a syringe and Hamilton polyethylene tubing are not available. The sector on the left is filled with a slightly larger volume of solvent (5–10 μL more than the sample volume).

3. Seal each filling hole with a red polyethylene gasket and brass screw, and hand-tighten. The filling holes may be double-gasketed; but do not over-tighten the brass screws because the cell housings will deform and not fit into the rotor.
4. Weigh the assembled cells and make sure that they are within 0.5 g of each other, including the counterbalance which must go in rotor hole #4. The counterbalance may be adjusted with brass weights that screw into the housing.
5. Install each cell with the screw-ring end up and the housing filling holes toward the center of the rotor, matching the scribe lines on the cells with those on the rotor.

3.4. Preliminary Scans for Determination of Experimental Parameters

1. Load the rotor straight down onto the centrifuge drive shaft. Install the monochromator assembly and start the vacuum pump.
2. Start the XL-A program and set the four-letter filename desired.
3. Perform a quick radial scan at 3000 rpm to check the positions of the menisci, using settings of 0.01 cm, 1 average, from 5.8 to 7.2 cm. Properly filled cells should give a scan similar to **Fig. 1**. Note radial positions A and B as the cell length to be scanned, and position C as a point near the center of the liquid column. A check of the lamp energy may be done at this time (*see* **Note 3**).
4. Perform a wavelength scan at a radius near position C, using step size of 1 nm, 1 average, from 180–650 nm. Choose the wavelength to use so that the absorbance falls between 0.2 and 0.35 for all cells. It is preferable to choose a wavelength at which the extinction coefficient of the sample is known.
5. Set up for data acquisition using the parameters determined in **steps 3** and **4** for radial scans, 0.001 step size, and 10 averages. Set the rotor speed and temperature determined in **Subheading 3.2.**
6. Perform an initial scan at experimental conditions. Overlay with the preliminary scan done in **step 3**, and check to see that the menisci fall directly on top of each other. A shift in the position of the meniscus means that a cell has leaked. In this case, the run must be stopped and the cell must be reloaded.

3.5. Data Acquisition

1. The XL-A measures absorbance as a function of radial position for each sample, and a separate data file is archived for each scan. The Beckman program names each primary data file with 11 characters, the first four of which are user-selected. (The remaining characters denote the cell number, the type of scan, the wavelength, and the number of the scan in an autoscan series.) A large number of files will be generated and we have found that deciding on a systematic four-character naming scheme before starting an experiment is very helpful.
2. Select **review active cell** to verify the desired conditions. Begin a single scan by selecting **start scan** or a series of scans automatically repeated at a set time interval by selecting **start autoscan**.
3. Overlay two scans of the same sample that were taken hours apart. Equilibrium has been achieved in that sample when the data points lie directly over one another.

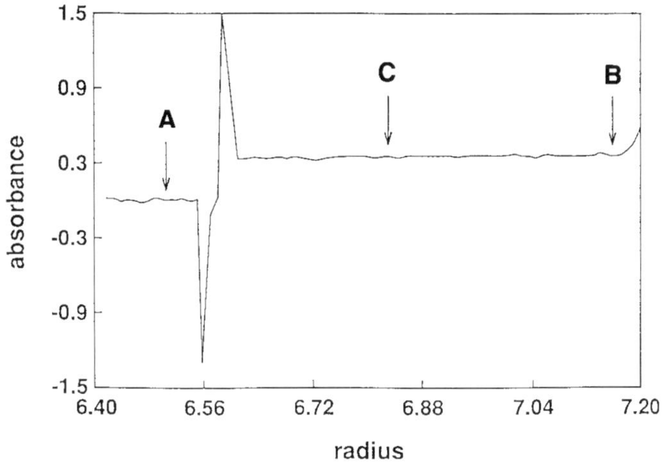

Fig. 1. A typical initial scan of a properly filled cell.

3.6. Data Analysis

1. Check all primary data files and choose the files to be analyzed. For each data set chosen, select the **S/ave subset** option. Type in up to eight filename characters, and the program will add the extension ".RA#" where # is 1, 2, or 3 to denote the cell providing the data.
2. Transport the saved files to the data analysis program. The objective is to "fit" the experimental data (concentration or absorbance and radial position) to a function whose parameters (solvent density, temperature, rotor speed, and so on) describe the system.
3. Each file should first be analyzed individually, assuming the simplest possible case. Previously determined values for $\bar{v}$, ρ, ω, R, and T are combined in the "A" term and used along with the known monomeric molecular weight in **Eq. 1.** Be sure to use rotor speed in radians/s (ω = rpm $\times$ 0.10472), R as 8.314E7, temperature T in degrees Kelvin, and R_b= 7.2 cm. Give initial guesses for c_b and ε, and ask the program to fit c_b and ε to the function using the specified data file. The program should generate the most probable values for the parameters c_b and ε, along with a statistical measure of precision and goodness of fit. The optimal values for the parameters are found by minimizing the sum of the squares of the differences between the experimental data and the fitted function. Convergence means that the curve-fitting process has reached a minimum value. Perform multiple analyses using different initial guesses for the parameters. If all converge to the same values, it is reasonable to assume that the values are unique. Always examine the values that result to be sure that they are physically meaningful.
4. Try fitting to additional models until satisfactory goodness of fit statistics are achieved, based on the maximum degree of association expected. A typical progression might be: monomer, dimer, trimer, monomer-dimer, monomer-trimer,

monomer-tetramer, monomer-hexamer, monomer-octamer, monomer-dimer-tetramer, monomer-dimer-tetramer-octamer, and so on. Comparison of the goodness of fit statistics will indicate the most probable equilibria.

5. More accurate values for association constants can be obtained by global fitting of several data files at once. Global parameters are those expected to have the same value for all data channels, such as molecular weights and ln k. Nonglobal parameters are those unique to each individual channel, such as the reference concentrations of monomer and rotor speed.

4. Notes

1. Cells that leak during ultracentrifugation are disastrous, causing losses of usable data and valuable sample material. Leaks can be minimized by taking care in cleaning and drying cell components. Soak windows and centerpieces in a solution of 2–3 mL of Beckman Solution 555 Rotor Cleaning Concentrate in 100 mL distilled water. Components should be inspected carefully while assembling cells and replaced when worn or damaged.
2. When possible, samples should be dialyzed against the solvent and a portion of the final dialysate reserved to use as the reference solution *(5)*.
3. The lamp output should be checked periodically by performing a wavelength scan from 180 to 650 nm on cell 4. A clean lamp will show peak maxima in the low UV around 230 nm and a minimum energy of 2000. Beckman recommends calling the service engineer to clean a dirty lamp.

References

1. Weast, R. C., ed, (1986) *CRC Handbook of Chemistry and Physics, 67th ed.*, CRC, Boca Raton, FL, p. F5.
2. Laue, T., Shah, B., Ridgeway, T., and Pelletier, S. (1992) Computer-aided interpretation of analytical sedimentation data for proteins, in *Analytical Ultracentrifugation in Biochemistry and Polymer Science*, The Royal Society of Chemistry, Cambridge.
3. Edelstein, S. J. and Schachman, H. K. (1967) *J. Biol. Chem.* **242,** 306–311.
4. Condino, J. (1992) The determination of molecular weights by sedimentation equilibrium. Technical Information DS-820, Beckman Instruments, Inc., Palo Alto, CA.
5. Condino, J. (1992) Sample preparation for analytical ultracentrifugation in the Optima XL-A. Technical Information, Beckman Instruments, Inc., Palo Alto, CA.

8

NMR Characterization of Amphipathic Helical Peptides

Xiaotang Wang and Kathleen M. Morden

1. Introduction

The remarkable advances and improvements in nuclear magnetic resonance (NMR) technology and methodology in recent years have made significant impact on the investigation of biological macromolecules, including amphipathic helical peptides. Through NMR studies, the three-dimensional structures of such peptides can now be obtained in solution at resolution levels comparable to those of single-crystal X-ray structures. NMR spectroscopy can provide a wealth of additional information about peptides in solution. For example, oligomerization, peptide-lipid interactions, and dynamics of the peptide can be investigated. The diversity of information obtainable from NMR data as well as the ability to study the peptides under conditions analogous to those found in vivo makes NMR spectroscopy increasingly attractive for the investigation of biological macromolecules.

Amphipathic helical peptides contain polar and nonpolar residues on opposing faces along the long axis of the helix. Because of this unique structural feature, the peptides have a strong tendency to self-associate and interact with detergent micelles and lipid bilayers. Many amphipathic helical peptides are also highly charged and thus the structural properties are strongly affected by solvent properties such as pH and counterion concentration. These characteristics provide challenges in applying the multidimensional NMR techniques necessary for resonance assignments and structure determination of amphipathic helical peptides. The amount of information that can be extracted from the NMR data is determined by the quality of the data and spectral properties such

From: *Methods in Molecular Biology, Vol. 78: Antibacterial Peptide Protocols*
Edited by: W. M. Shafer, Humana Press Inc., Totowa, NJ

as resonance linewidths and chemical shift dispersion. Whereas amphipathic helical peptides can be challenging NMR samples, NMR is uniquely suited for investigating the structure and dynamics of these peptides.

The NMR observable nuclei that are most relevant to the investigation of amphipathic helical peptides are ^{1}H, ^{2}H, ^{13}C, and ^{15}N. Among these nuclei, the proton has the highest natural abundance and is the most sensitive to detection. Thus the proton is, by far, the most extensively used nucleus. Therefore, we will focus our discussion on the application of ^{1}H NMR in the characterization of amphipathic helical peptides. We also limit the scope of our discussion to the solution state although solid state NMR techniques have also been used to study amphipathic helical peptides.

This chapter describes the optimization and application of NMR techniques generally used in the investigation of amphipathic helical peptides. We have put our emphasis on the information content and applications rather than the physical principles behind the experimental methods. Many of the techniques discussed here are adapted from protein studies. Overviews and more detailed descriptions of these protein applications can be found in several recent reviews or the references therein ***(1–6)***.

2. Sample Preparation

We begin with some of the most common problems encountered in the preparation of an NMR sample, since this seemingly simple subject is actually the determining factor for obtaining well-resolved, reproducible, and informative spectra from the NMR spectrometer. The importance of proper handling of the sample cannot be overemphasized; even the expensive and sophisticated spectrometer or computer equipment cannot compensate for a poorly prepared sample. Careful handling of the sample can save hours of spectrometer time and, even in some cases, the peptide.

2.1. Requirements for the Peptide

2.1.1. Stability

NMR experiments often require that the sample remain at room temperature or higher for many days. Thus the peptide must be stable over this time frame. This does not present a severe problem as most amphipathic helical peptides are fairly stable. However, this does not preclude taking precautions in handling the peptide samples. Sample conditions that promote peptide degradation, such as air oxidation, microbial contamination, and hydrolytic breakdown, should be avoided to maximize the lifetime of the sample in the NMR tube. Air oxidation can be minimized by removing the dissolved O_2 using the freeze–thaw method or cycles of evacuating the gas from the sample in the NMR tube followed by filling with a dry inert gas such as argon. Microbial contamination during sample storage of amphipathic helical peptides is not usually a serious problem because many of the peptides are antibacterial. Even for peptides that lack antibacterial activity, the solvents (D_2O, organic solvents, and detergents) used for the peptide studies are often efficient in suppressing bacterial growth. However, it is not advisable to store samples in H_2O for a long period of time. For peptides that lack antimicrobial activity and are to be stored in water for an extended period of time, trace amounts of azide (<50 μ*M*) or toluene-d_8 (<2% v/v) can be added to the sample to suppress bacterial growth. Hydrolytic loss of the peptide can be minimized by avoiding prolonged exposure of the sample to extremes of pH and temperature. A 1D ^1H NMR spectrum of the sample should be recorded before and after long data acquisition. The appearance of new resonances with fractional intensities may indicate degradation of the peptide.

2.1.2. Molecular Weight

In general, the complexity of the spectrum and the linewidth of the resonances increase with increasing molecular weight. Large molecules may require multidimensional multinuclear techniques or can result in intractable spectra and cancellation of crosspeaks in multidimensional experiments. Fortunately, NMR techniques have continuously increased the upper size limit of molecules that can be studied. Molecules with molecular weights up to 30 kDa have been successfully investigated using heteronuclear multidimensional NMR spectroscopy ***(2,3,5,7)***. Molecules with molecular weights below 10,000 are fairly routine and often only require homonuclear techniques. The molecular weights of most amphipathic helical peptides (less than 50 residues) fall into the favorable range for detailed structural determination by NMR spectroscopy without requiring isotopic labeling ***(8)***.

2.1.3. Quantity

Although the potential information content of NMR spectroscopy is very high, NMR is many orders of magnitude less sensitive than most spectrophotometric techniques. This leads to the requirement of large amounts of sample. For routine 1D data acquisition, a 400- to 600-μL solution with a concentration of at least 1 m*M* peptide is required. More dilute samples can be studied, but will require extensive signal averaging and will therefore limit the studies to 1D experiments. For multidimensional experiments, concentrations of >5 m*M* are recommended. If unlimited quantities of peptide are available, very high concentrations may be used while keeping in mind that aggregation may occur, leading to increased resonance line width. Aggregation can be monitored by looking for changes in the line width or chemical shift in a 1D spectrum as a function of peptide concentration. Concentration independent line width and chemical shift do not necessarily assure the absence of oligomerization, as it is possible that the peptide is present as aggregated forms throughout the experimental concentration range. It is highly recommended that the aggregation state of the peptide also be monitored using complementary techniques such as gel electrophoresis, size exclusion chromatography, or analytical ultracentrifugation. If availability of material is very limited, the sample volume can be reduced by using specially designed, commercially available NMR tubes (Shigemi, Tokyo, Japan), thus reducing the quantity of the sample needed. These tubes contain inserts made of material that matches the magnetic susceptibility of water, and thus will not compromise the quality of the spectrum. More spectacular improvement on minimizing the required sample volume may be forthcoming with the development of specially designed rf coils which require only a few nanoliters of solvent ***(9)***.

2.1.4. Purity

Assuring the purity of the peptide sample prior to conducting NMR experiments is critical. The presence of intense signals from impurities either requires the use of special techniques to obtain usable spectra or simply results in uninterpretable spectra. Techniques such as analytical HPLC and mass spectroscopy are typically used to assess the purity of peptide samples.

2.2. The NMR Tube

2.2.1. Types of Tubes

A wide range of NMR tubes with varying specifications are commercially available from a number of vendors. For most biological studies, it is worthwhile to invest in the highest quality NMR tubes. Lower quality NMR tubes can cause a variety of spectral anomalies resulting in broader resonances and can limit the attainable field homogeneity. As mentioned previously, there are also NMR tubes specifically designed to handle minimal sample volume.

2.2.2. Preparing the Tube

NMR tubes straight from their shipping packet may contain chemicals associated with their manufacturing and should never be used without thorough cleaning. A good way to clean NMR tubes is to soak them in a nonphosphate detergent solution for a period of at least 24 h and then rinse exhaustively with distilled water. Rinsing the NMR tubes can be efficiently accomplished using a commercially available NMR tube washer. Organic solvents, such as acetone, are not recommended for cleaning NMR tubes used for biological samples as they may contain impurities that will be deposited on the inside surface of the tube.

For samples dissolved in solvents other than H_2O, adequate drying of the NMR tube is highly desired for minimizing interference from water. A simple and practical way to dry an NMR tube is to blow a stream of filtered dry nitrogen gas through a pipet against the bottom of the tube or by keeping the tube inverted in a dust-free oven at no more than 50°C. Use of very high temperatures is not recommended, as it degrades the high specifications of the tube.

2.3. Optimizing Sample Conditions

The sample conditions discussed here consist of the solvent used to dissolve the peptide, the buffer (particularly the counterions), the pH, and the temperature. These factors can influence the conformation of the peptide as well as the quality of the NMR spectra. It is difficult to know *a priori* the best conditions for a given peptide. Therefore, it is necessary to try different combinations of solvent, counterion, pH, and temperature in order to optimize the quality of the spectra.

2.3.1. Solvent

Solvents including pure aqueous, mixed aqueous/organic, pure organic, and aqueous detergent micelles, have been widely employed in the literature for studying amphipathic helical peptides. Pure aqueous solution is often an ineffective solvent as most amphipathic helical peptides contain hydrophobic groups and are either sparingly soluble or aggregate, leading to broad and uninterpretable NMR spectra ***(10)***. Even for peptides that can be solubilized in pure aqueous solution, they often exist as unstructured random coils. This lack of secondary structure is reflected by a limited spectral dispersion in the 1D NMR spectrum as shown in **Fig. 1A**.

Organic solvents (either pure or mixed with aqueous solution) and detergent micelles are effective solubilization media for NMR investigations of amphipathic helical peptides. Ordered peptide structures have been observed in both organic solvent and detergent micelles and are reflected by a dispersion of the amide proton resonances (**Fig. 1B**) ***(11,12)***.

Organic solvents, such as methanol ***(13)***, trifluoroethanol ***(14,15)***, and hexafluoroisopropanol ***(16)***, mixed in different ratios with water have several practical advantages. These organic solvents are readily available at reasonable cost from commercial sources in either fully or partially deuterated forms. The low viscosity of these organic solutions gives rise to short correlation times for the peptide and therefore narrow NMR resonances.

Detergent micelles are another effective solvent for solubilizing amphipathic peptides and are considered a better mimic of the molecular environment of membranes compared to organic solvent. Monomer units of the detergents consisting of lipids with hydrophilic head groups will aggregate into micelles at concentrations above the critical micelle concentration (cmc). The polar head group of a micellar detergent interacts with the aqueous environment, whereas the hydrophobic tail forms a central water-excluding core.

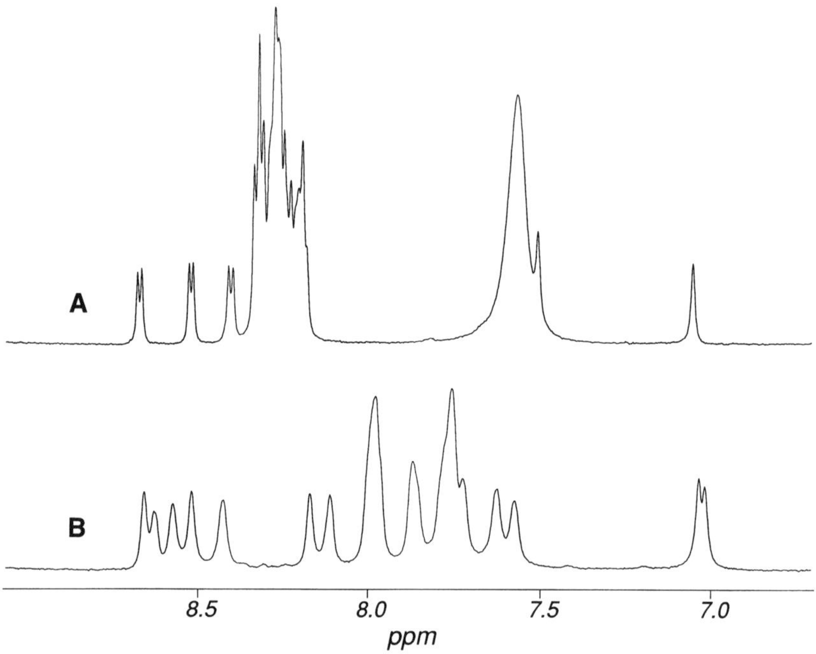

Fig. 1. The amide region of the proton NMR spectrum of the amphipathic helical peptide $(KLAKKLA)_3$ collected at 500 MHz in 90% H_2O/10% D_2O at 25°C, pH 3.3 in 2.5 m*M* phosphate buffer. **(A)** aqueous solution and **(B)** 15% (v/v) HFIP.

Sodium dodecyl sulfate (SDS) and dodecylphosphocholine (DPC) are the most commonly used detergents for solution NMR studies of amphipathic helical peptides. Both detergents form stable micelles with small aggregation numbers ***(17,18)***; thus the peptide bound to micelles will retain a relatively small correlation time and still result in narrow NMR resonances. Both SDS and DPC are commercially available in perdeuterated form, which minimizes background from detergent proton resonances. With gradient spectroscopy and regioselective excitation methods, nondeuterated detergents can also be used ***(19)***.

The detergent:peptide ratio is one of the most important factors that affect the quality of the NMR spectra of micelle-bound peptides as shown in **Fig. 2**. Detergent concentrations much higher than the cmc are usually needed to obtain optimum spectra. However, concentrated detergent solutions are usually very viscous and it is often necessary to compromise. 1D spectra of the peptide can be obtained with increasing concentrations of detergent to determine the detergent:peptide ratio that produces narrow line widths and maximum spectral resolution as shown in **Fig. 2**. Usually, the detergent:peptide ratio must be at least 100:1 to obtain the narrowest resonances. If the peptide is

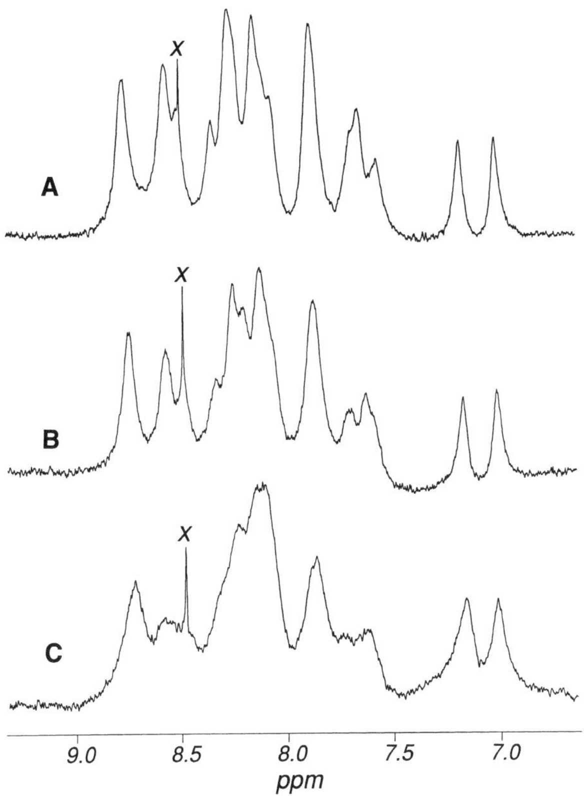

Fig. 2. The amide region of the proton NMR spectrum of $(KLAKKLA)_3$ collected at 500 MHz in 90% H_2O/10% D_2O at 25°C, pH 7.0 in 10 m*M* phosphate. Molar ratios of SDS:peptide are **(A)** 200, **(B)** 100, and **(C)** 50. X indicates an impurity.

isotopically labeled, the effect of detergent concentration can also be monitored using a heteronuclear multiple quantum coherence spectroscopy (HMQC) experiment ***(20)***.

A disadvantage of detergent micelles is that the bound detergent adds considerably to the effective molecular weight of the peptide, resulting in slow isotropic reorientation in solution. This leads to broader line widths compared to aqueous solution and efficient spin diffusion that may limit the applicability of many homonuclear solution NMR experiments used in resonance assignment.

2.3.2. pH

The pH of a peptide sample can affect the NMR spectrum of the peptide by changing either the amide proton exchange rates or the ionization state of the side chains or both. Acidic pH, 3.0–5.0, is often used for NMR studies because the amide proton exchange rates are generally suppressed at acidic pH *(18)*. However, it should be noted that side chains from aspartic acid, glutamic acid, and histidine are protonated at acidic pH.

Special care is needed in adjusting the pH of small volume NMR samples to avoid local extremes of pH that may be detrimental to the peptides. The NMR sample should be removed from the tube to a microcentrifuge tube or glass vial using a long tip glass pipet or syringe fitted with a long Teflon needle. Adjustment of pH is carried out by placing a few microliters of acid (or base) on the side of the vial, and vortexing to allow thorough and rapid mixing of the sample with the acid (or base). A detailed procedure for adjusting the pH of biological NMR samples is available *(21)*. The pH dependence of the counterion concentration must also be considered in selecting a proper buffer system for a particular peptide, which is discussed further in the **Subheading 2.3.3.**

2.3.3. Counterions: Ionic Strength

Because most of the amphipathic helical peptides are highly charged, variation of the concentration of salts (including buffer ions) can affect the structure and conformation of the peptide, leading to dramatically different NMR spectra. **Figure 3** shows the effect of salt (sodium sulfate) concentrations on the amide region NMR spectra of a highly charged peptide in aqueous solution. The presence of salt screens the charge repulsion between peptide molecules and, therefore, affects the structure and aggregation state of the peptide which in turn affects the NMR spectrum. At pH 7.0 and low sulfate concentration, the peptide is largely random coil and amide protons exchange rapidly with the solvent. At high sulfate concentration, the peptide forms an α-helix and is aggregated, protecting the amide protons from exchange. For peptides dissolved in detergent micelles, ionic strength also affects the cmc and aggregation number of the detergent *(22)*. In practice, both ionic strength and the type of counterion can be varied to optimize the NMR spectrum. This can be achieved by monitoring the 1D spectrum of the peptide as ionic strength (or ion type) is changed. Consequences may vary for different peptides as ionic strength is changed. If the peptide is charged, increasing the ionic strength screens the charge repulsion and favors the formation of aggregated forms of the peptide leading to broader NMR resonances. However, if the peptide contains salt bridges, increasing the ionic strength can stabilize the monomeric state of the peptide, giving rise to sharper resonance lines. A detailed description of optimizing the NMR spectra by varying ionic strength can be found in the literature *(21)*.

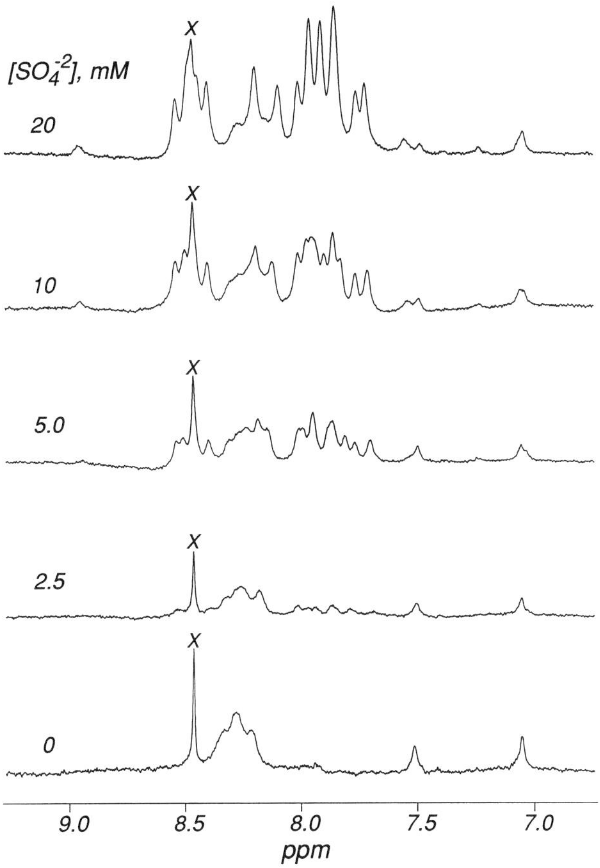

Fig. 3. The amide region of the proton NMR spectrum of $(KLAKKLA)_3$ collected at 500 MHz in 1.25 m*M* phosphate, 90% H_2O/10% D_2O, 25°C, pH 7.0, as a function of sulfate concentration (shown at the left of each trace of the spectrum). X indicates an impurity.

Although the procedure was developed for other biological macromolecules, it applies equally well for amphipathic helical peptides.

Some buffer systems contain ionizable groups that can act as counterions for the charged peptide molecules. Therefore, changing pH will also change the counterion concentration from the buffer. The selection of a buffer system also depends on the range of pH to be employed in studying the peptide. Phosphate is an example of a buffer that has a wide pH range and is often used in studies of biological molecules. The concentration of monovalent, divalent, and trivalent anions in the phosphate buffer will vary as a function of pH and thus caution should be exercised in using this buffer with highly positively charged peptides. This problem can be avoided by using a buffer devoid of these ionizable counterions or by providing another source of counterions such as sulfate (**Fig. 3**).

2.3.4. Temperature

The NMR spectrum of most biological samples, including amphipathic helical peptides, is strongly temperature dependent due to conformational equilibria and exchange phenomena. In general, the resolution of the NMR spectrum will increase with increasing temperature as shown in **Fig. 4**. This is owing to the decrease in the correlation time of the sample. For some peptide samples in SDS micelles, it is imperative that the experiment be carried out at elevated temperatures in order to obtain observable 2D NMR spectra. This is mainly owing to the increased effective molecular weight of the micelle-bound peptide that gives broad resonances at room temperature. The increased linewidth can cause cancellation of the crosspeaks in correlated spectroscopy (COSY) experiments. In practice, the thermal stability of the peptide secondary structure and the purpose of the experiment must also be considered in addition to spectral resolution in determining the best temperature. If, for example, the amide proton exchange kinetics are to be studied, high temperature should be avoided as increases in temperature result in dramatic increases in the exchange rates, making measurement difficult. A series of 1D spectra run at several different temperatures, such as those shown in **Fig. 4**, is a very efficient way to determine the optimum temperature for a particular sample.

2.4. Sample Handling

Samples for high resolution NMR should be free of particulate matter, colloids, emulsions, and so on. This can be achieved by filtering the sample solution directly into the NMR tube. A small plug of prerinsed fresh cotton wool at the neck of a Pasteur pipet is an effective filter that is usually available in any research laboratory. Another convenient and efficient way to filter an NMR sample is to use a 0.45-µm syringe filter. Such syringe filters are compact and have negligible retention volume, which is highly desirable for filtration of small volume samples. It is extremely important that these filters be washed before use as they often contain glycerol (as a preservative), which will contaminate the peptide sample. Centrifugation of the sample before placing it into the tube also removes some suspended matter but this is not as efficient as filtration. It is highly recommended that gloves are worn in handling very dilute biological samples in order to avoid contamination, which can result in signals at unpropitious chemical shifts for peptide samples (1.4 and 4 ppm) ***(23)***.

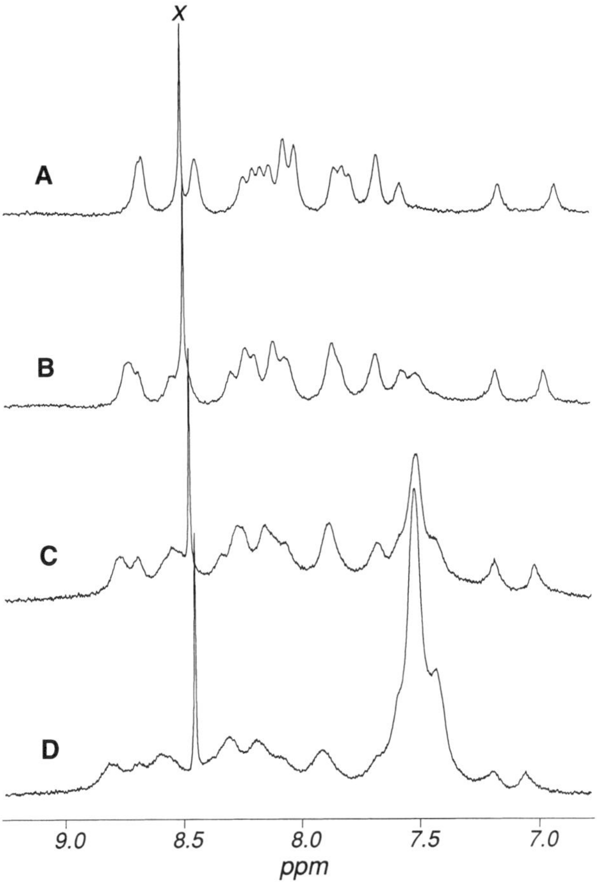

Fig. 4. The amide region of the proton NMR spectrum of $(KLAKKLA)_3$ collected at 500 MHz in 90% H_2O/10% D_2O, 10 m*M* phosphate, 400 m*M* SDS, pH 4.3. **(A)** 45°C, **(B)** 35°C, **(C)** 25°C, **(D)** 15°C. X indicates an impurity.

3. One-Dimensional NMR Spectroscopy

One of the significant parameters that can be derived from 1D NMR experiments is the chemical shift, particularly for the NH and CαH from the backbone of the peptide. Even without resonance assignments, the distribution of chemical shifts observed in a 1D NMR experiment can be used as a crude estimate of the presence of ordered structures. The chemical shifts for protons in the common amino acid residues in random coil polypeptides have been extensively studied and well characterized ***(11,24,25)*** (**Fig. 5**). In the random coil, resonances from the amide and α-carbon protons fall within a small range of chemical shifts. However, formation of secondary structure, particularly

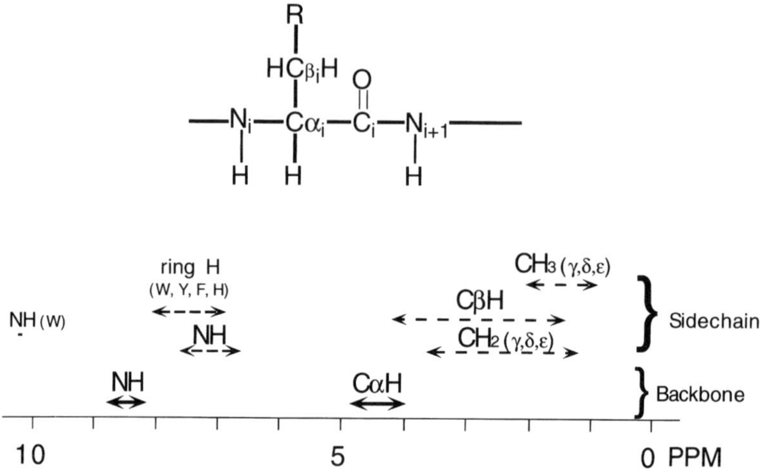

Fig. 5. Schematic of the peptide backbone (top) and the range of ^{1}H chemical shifts for common amino acids in a random coil conformation (bottom).

α-helix, results in dispersion of the chemical shift (**Fig. 1B**). 1D NMR experiments can also provide information about whether the peptide aggregates (by monitoring the concentration dependence of the NMR line widths), whether the secondary structure is stable as a function of temperature and pH, and the magnitude of through-bond coupling constants. Most importantly, the 1D spectrum provides insight as to whether the sample is suitable for extensive NMR studies and allows for optimization of the experimental conditions for a thorough NMR investigation of the peptide.

3.1. Acquiring Spectra

Although performing 1D NMR experiments requires no complex spectrometer set-up, it is highly recommended that you consult the NMR facility manager before you get on the spectrometer if you are new to NMR. For all of the experiments described below, we have assumed some fundamental knowledge of NMR spectroscopy, such as 90° pulse, field lock, probe tuning, magnet shimming, relaxation delay, repetition rate, and so on.

Setting up the spectrometer for the collection of 1D spectra involves the following fundamental operations: establishing the field lock, tuning the probe, shimming the magnet, calibrating the rf pulse length, and optimizing the acquisition parameters. Tuning, shimming, and the 90° pulse length are strongly sample dependent and can be affected by factors such as solvent type, ionic strength, temperature, sample volume, and sample position. Therefore, if any of these parameters are changed it is necessary to readjust the tuning and shimming, and recalibrate the 90° pulse length. The accurate determination of the 90° pulse length is even more important for the collection of 2D NMR data.

The spectral width, transmitter offset, receiver gain, and relaxation delay are the parameters most frequently adjusted in optimizing the quality of a 1D spectrum.

3.1.1. Spectral Width

It is recommended that a spectral width of approx 50% larger than the spectral region of interest be used in order to alleviate severe baseline distortions. This is particularly important for 2D experiments.

3.1.2. Receiver Gain

For most spectrometers, the receiver gain can be set automatically; however, this should be used with caution. Too large a receiver gain creates baseline distortion in the spectrum especially near intense signals.

3.1.3. Transmitter Offset

The transmitter offset is recommended to be set at the same frequency as the most intense peak in the spectrum (usually the solvent peak for biological samples) in order to minimize spectral artifacts.

3.1.4. Relaxation Delay

The relaxation delay should be set to at least $1.25 \times T_1$, in which T_1 is the longitudinal relaxation time of the nuclei of interest. For most peptide samples, a relaxation delay of 2 s should satisfy this requirement. However, measurement of the T_1 relaxation time is strongly recommended before collecting extensive data.

3.2. Observing Resonances from the Labile Protons

For NMR studies of amphipathic helical peptides in aqueous solution, spectra must be obtained in H_2O in order to observe the exchangeable proton resonances, particularly those from the amide protons. In such cases, the intense solvent signal must be suppressed in order to detect the weaker peptide signals. This can be achieved using several different methods ***(26)***:

1. Preirradiation of the solvent resonance with a low-power, frequency-selective, continuous-wave radiofrequency pulse ***(27)***.
2. Selective excitation (excluding the solvent resonance) with a series of either long, weak (soft) pulses or short, strong (hard) pulses separated by delays ***(28)***.
3. A gradient echo pulse sequence such as WATERGATE ***(29)*** used for single quantum experiments such as nuclear Overhauser effect spectroscopy (NOESY).

Preirradiation of the solvent is easy to implement; however, if the amide protons are in rapid exchange with solvent protons, this will result in the loss of amide proton resonances. Preirradiation of the H_2O signal can also cause loss or reduction of the intensity of resonances nearby the H_2O resonance, most notably the α-proton resonances. The problems with rapid exchange of the amide protons are alleviated by using selective excitation, although problems still exist with reduced excitation near the solvent resonance. There can also be problems with baseline distortion using this approach. Implementation of the selective pulse sequences is more challenging than preirradiation, but has become a standard experiment on most spectrometers. Assuming the necessary hardware is available, the gradient method is the best approach for observing both the amide and alpha protons with minimal distortion.

4. Two-Dimensional NMR Spectroscopy

NMR studies of all biological macromolecules, including amphipathic helical peptides, depend heavily on the successful collection of a series of 2D or higher dimensional NMR spectra. A 2D experiment consists of a preparation period, an evolution period (t_1), a mixing period, and an observation period (t_2). The experiment is a series of 1D experiments collected at increasing intervals of the evolution period, thus sampling the signal as a function of t_1. The result is a data matrix $S(t_1, t_2)$. The data is Fourier transformed with respect to t_1 and t_2 to give a 2D spectrum that is a function of two frequency variables S(F1, F2). In a homonuclear 2D NMR spectrum, the peaks on the diagonal contain the chemical shift information as would be observed in a 1D spectrum and the peaks off the diagonal (crosspeaks) indicate the interactions between protons. Interactions that are commonly probed using 2D experiments are chemical exchange, nuclear Overhauser effect (NOE) and J-coupling. The NOE is a particularly important interaction for probing the structures of peptides because of the strong (r_{ij}^{-6}) distance dependence of the interaction. As will be discussed later, one should always keep in mind that if there are multiple solution conformations in rapid exchange on the NMR time scale, the parameters measured in the NMR experiment are an average *(8)*. The average will not just be a linear combination weighted by the population of each conformation. For example, the NOE is weighted as r_{ij}^{-6}, thus conformations in which two protons are close to each other will be weighted more heavily than those where the protons are far apart.

4.1. Acquiring 2D Data

Many of the procedures and parameters involved in the setup of 2D NMR experiments are identical to those involved in the setup of 1D experiments, which were summarized in **Subheading 3.1.** Setting the receiver gain for a 2D experiments can require special care, especially for experiments using solvent suppression, because the signal strength can increase as t_1 increases. This leads to digitizer overload in the later stages of the experiment. The setting for receiver gain can be determined using the 2D pulse sequence to run trial 1D experiments with several t_1 values ranging from the minimum to the maximum value. The largest receiver gain acceptable for all of the tested t_1 values should be selected for the actual 2D experiment.

Several parameters are unique to 2D (or higher order) experiments and are discussed below.

4.1.1. The F1 Spectral Width

Ideally, the spectral width in F1 is set to the minimum possible value so as to maximize sensitivity and minimize the number of acquisitions necessary in the F1 dimension. It is typical to set the spectral width in F1 to the same value or to a value related by a factor of two to the spectral width in F2. Setting the spectral width will automatically determine the time interval that will be used to sample during t_1.

4.1.2. The Number of Experiments in F1

The number of experiments in F1 is set to give the maximum spectral resolution while keeping the total experimental time within a reasonable time frame (usually less than 40 h for a 2D experiment). There is a tradeoff in the number of scans per acquisition (*see* **Subheading 4.1.3.**) and the number of experiments that can be taken.

4.1.3. The Number of Scans per Acquisition

The rule of thumb for the number of scans to be averaged is that the total number of scans accumulated in the experiment must give an adequate signal-to-noise ratio spectrum. This can easily be achieved by collecting a spectrum in the 1D mode using identical acquisition parameters of the 2D pulse sequence. Also keep in mind that the number of scans must be set in multiples of the phase cycle.

4.2. Resonance Assignments

Several approaches have been developed for the assignment of resonances from a protein or peptide using homonuclear NMR techniques, the sequential method ***(30,31)*** and the main chain-directed method ***(32)***. The sequential resonance assignment procedure is briefly discussed here.

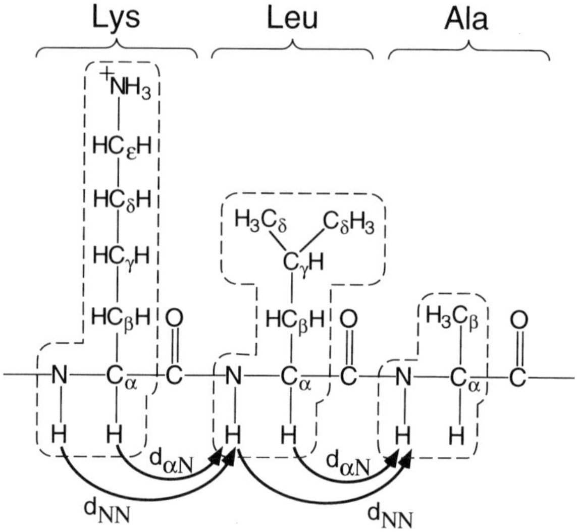

Fig. 6. Schematic of the spin systems (inside dotted lines) of individual amino acid residues in a polypeptide segment. The sequential amide-amide and alpha-amide NOEs are shown with thick arrows.

The sequential assignment method can be divided into several steps. The first step is the identification of distinct spin systems belonging to a particular amino acid residue or a class of residues. A spin system is a group of spins connected by scalar (through-bond) spin–spin couplings, as shown schematically in **Fig. 6**. The identification of spin systems is achieved using COSY type experiments, such as double quantum filtered (DQF) COSY, RELAY COSY, and total correlation spectroscopy (TOCSY, also known as homonuclear Hartmann-Hahn, HOHAHA, spectroscopy). If the experiments are conducted in fully deuterated solvent, a TOCSY experiment will identify through-bond connectivities of the alpha proton with the side chain protons in the same residue. The total correlation is achieved through isotropic mixing created by composite pulses, typically MLEV or DIPSI sequences. The TOCSY experiment for a relatively small peptide is typically obtained using an isotropic mixing time of 25–70 ms (with a low power 90° pulse length of 25–30 μs). TOCSY experiments with coaddition of several different mixing times can also be useful *(33)*. The spin system can be extended to include the amide proton by collecting a COSY spectrum in H_2O and using the fingerprint region to correlate the α-proton and amide proton chemical shifts. The resolution of the amide proton resonances is often better than for the α-protons. Conducting the TOCSY experiment in H_2O (90% H_2O/10% D_2O) allows for correlation of the

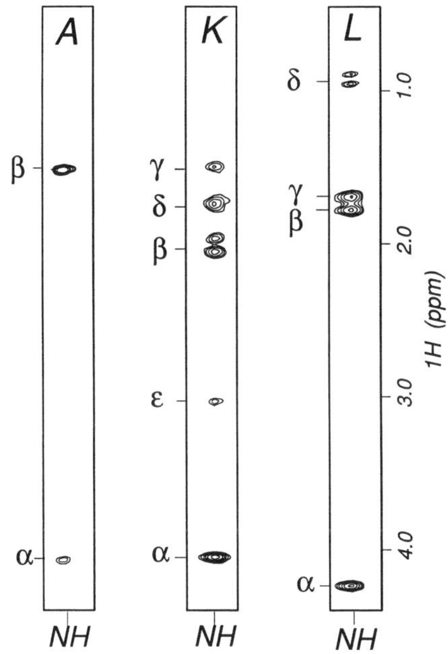

Fig. 7. Expansions of several regions from a TOCSY experiment showing correlations between the amide, α, and side-chain protons. Assignments of the α and side-chain proton resonances are indicated on the left of each expansion. The identity of the residue is indicated at the top of each expansion (*A*, alanine; *K*, lysine; *L*, leucine). The sample was 6 m*M* peptide, $(KLAKKLA)_3$, in 85% H_2O/15% HFIP, 2.5 m*M* phosphate, 25°C and pH 3.06.

amide proton with the alpha proton and the side chain protons within a residue or spin system while taking advantage of the improved resolution (**Fig. 7**).

The second step of the assignment procedure requires independent knowledge of the sequence of the peptide or protein. This step assigns the previously identified spin systems to a specific residue within the amino acid sequence using NOEs between sequential residues (**Figs. 6** and **8**). Detailed descriptions of assigning resonances using the sequential resonance assignment method can be found elsewhere ***(4,24,34)***.

5. Structural Characterization

Once the sequence-specific resonance assignments are obtained, the solution structure of an amphipathic helical peptide can be investigated using NOE, spin–spin coupling constants, and amide proton exchange rates. These parameters can be used qualitatively to determine a low resolution structure or more

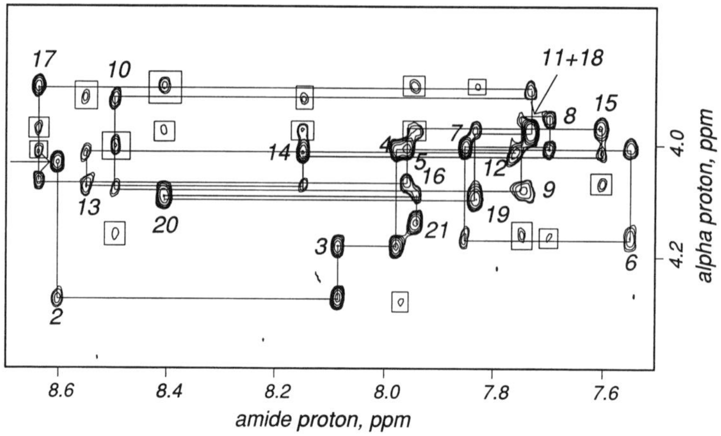

Fig. 8. Expansion of the fingerprint region of the NOESY spectrum. Intraresidue crosspeaks between amide and alpha protons are labeled with numbers indicating the position in the peptide sequence. Intraresidue and sequential interresidue crosspeaks are connected by a solid line starting at the sequential crosspeak indicated by the arrow on the left. Boxed crosspeaks are from medium-range NOEs. Conditions were the same as for **Fig. 7**.

quantitatively, in conjunction with molecular dynamics calculations, to obtain a higher resolution structure.

5.1. Identification of Secondary Structures

The application of NMR spectroscopy to the determination of protein and peptide structures is now well established. Extensive statistical data dealing with the qualitative relationship between regular secondary structures and various NMR parameters are available from several sources ***(11,12,24,34)***. As a result, secondary structures can now be identified on the basis of a number of readily measurable NMR parameters associated with backbone resonances, such as chemical shift values of the amide protons and α-protons, amide proton exchange rates, NH-αH coupling constants, and short-, medium-, and long-range NOEs.

5.1.1. The Chemical Shifts of the Amide and α-Protons

The relationship between peptide secondary structures and the chemical shifts of the amide and α-protons has been extensively studied and well documented ***(11,12)***. In general, helical structures have αH chemical shifts upfield from the average value for the extended random chain conformations (**Figs. 5**

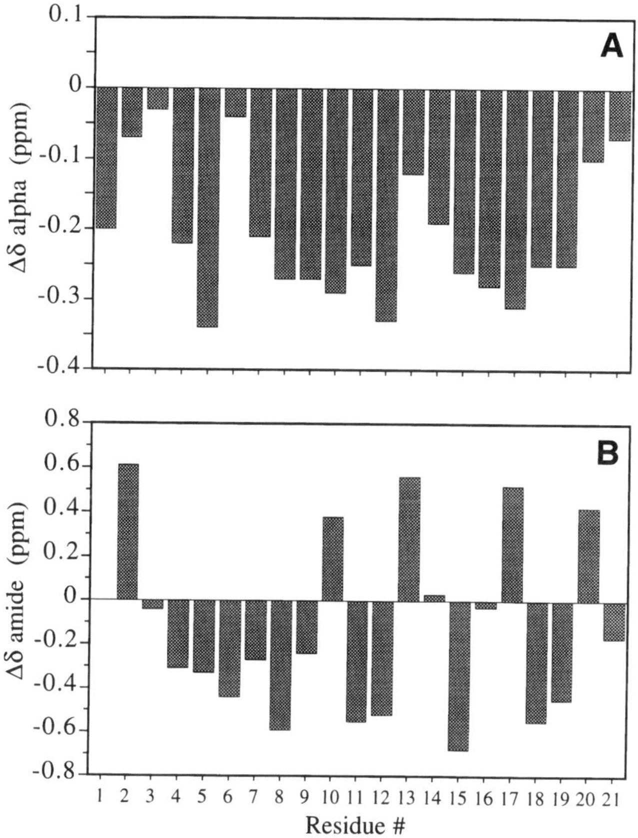

Fig. 9. Chemical shift difference plots (Δδ = δactual – δrandom coil) for **(A)** α proton resonances and **(B)** amide proton resonances from the helical peptide $(KLAKKLA)_3$. Random coil chemical shifts are taken from Wishart et al. ***(12)***.

and **9**). For helical peptides, the change in chemical shift of the amide protons relative to the random coil conformation follows a periodicity analogous to the helix periodicity (**Fig. 9B**) ***(35–38)***.

5.1.2. Amide Proton Exchange Rates

The formation of hydrogen bonds between backbone amide protons and backbone carbonyl oxygens is a feature of several secondary structures. For an α-helix, such hydrogen bonding occurs between the carbonyl oxygen of the i residue and the amide hydrogen of the i + 4 residue. Hydrogen bonding can be inferred by monitoring the rate at which the amide protons exchange with

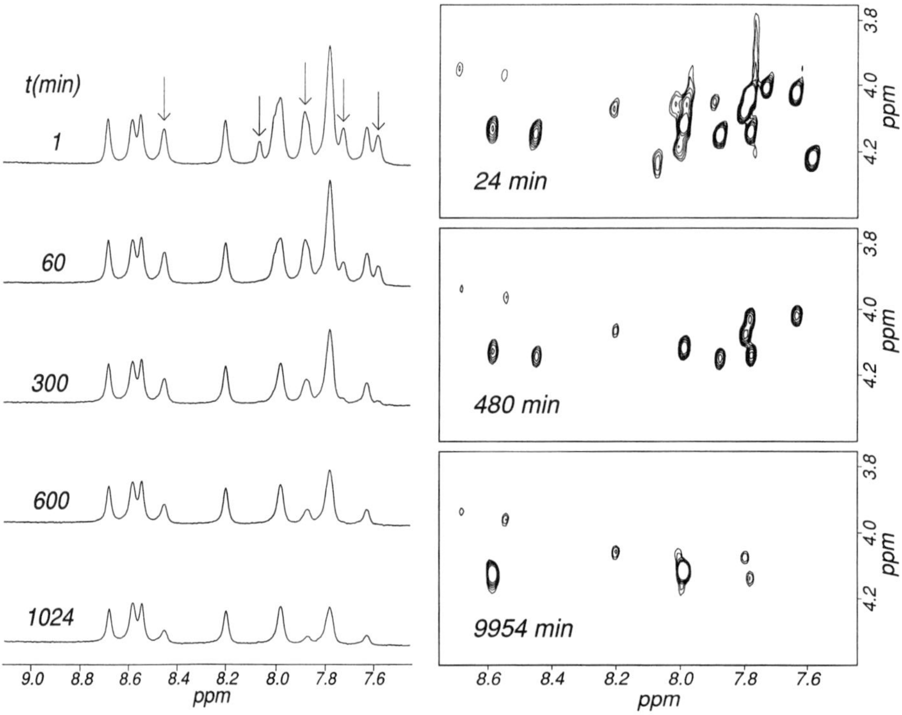

Fig. 10. One-dimensional (left) and TOCSY (right) data of the helical peptide $(KLAKKLA)_3$ in 85%D_2O/15% HFIP(d2) as a function of time (indicated on spectra) after dissolving in D_2O. The arrows indicate resonances from rapidly exchanging amide protons.

the solvent. The exchange rate constants for individual amide protons can be determined using either 1D or 2D experiments. The most common 2D proton experiment used is the TOCSY experiment because it provides enhanced spectral resolution and is a reasonably sensitive experiment. For isotopically labeled peptides (15N), the HMQC experiment can also be used. The peptide sample is lyophilized and then dissolved in D_2O and data collection is immediately begun. The exchange experiments are usually conducted at acidic pH, where the intrinsic exchange of the amide proton is at a minimum. The amount of acid or base needed to adjust the pH of a given sample for the experiment in D_2O must be precalibrated and the actual pH measured after data acquisition because data collection must begin immediately after dissolving in D_2O. Data from 1D or 2D experiments is collected as a function of time after dissolving in D_2O (**Fig. 10**). The natural logarithm of the intensities of the 1D amide resonances or the volumes of the α-amide crosspeaks from the TOCSY experiment

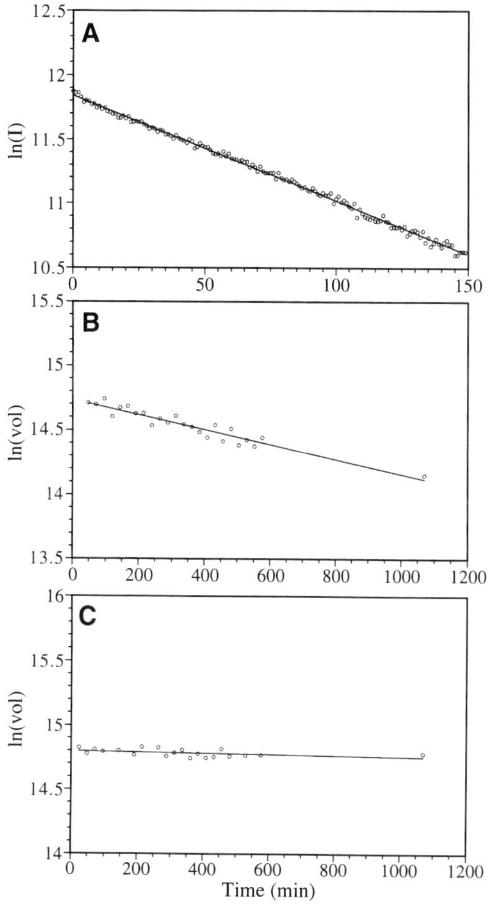

Fig. 11. Natural logarithm of the intensity of the resonance or volume of the crosspeak as a function of time after dissolving in D_2O. The line is determined with a linear least squares fit to the data points; the slope of the line is $-k_{ex}$. **(A)** Fast exchanging amide proton, $k_{ex} = 8.2 \times 10^{-3}$ min^{-1}; **(B)** intermediate exchanging amide proton, $k_{ex} = 5.7 \times 10^{-4}$ min^{-1}; **(C)** slowly exchanging amide proton, $k_{ex} = 4.3 \times 10^{-5}$ min^{-1}.

are then plotted as a function of time and the exchange rate constant (k_{ex}) is determined from the slope (**Fig. 11**).

The observation of slowly exchanging amide protons indicates the presence of secondary structure. The intrinsic or random coil exchange rate constant (k_{rc}) for the amide proton can be calculated as a function of the identity of the adjacent amino acid in the sequence of the peptide, the pH and the temperature of the sample ***(39,40)***. The ratio of the observed exchange rate constant to the intrinsic exchange rate constant can then be calculated for individual amide protons; the logarithm of the ratio is called the protection factor. When combined with sequence-specific resonance assignments, the protection factors can be plotted vs the amino acid sequence, providing additional information about the structural features of the peptide. For the synthetic amphipathic helical peptide $(KLAKKLA)_3$, two features are apparent from the plot shown in **Fig. 12**. The amide proton exchange rate constants for residues 5–20 are suppressed relative to the random coil exchange rate constants by factors of 5–1000, indicating the presence of hydrogen bonding. The amide protons show a periodic variation in exchange rate constants indicating the presence of helical secondary structure and characteristic of the formation of dimer or higher order aggregates of the peptide ***(41)***.

5.1.3. J Coupling Constants

The three-bond coupling constants between amide protons and the α-protons of the same amino acid residue, $3J_{NH\text{-}\alpha H}$, are very useful as a diagnostic of the secondary structure. It has been found for a large number of peptides and proteins that helical secondary structures have $3J_{NH\text{-}\alpha H} < 6$ Hz, whereas β-sheets have $3J_{NH\text{-}\alpha H} > 8$ Hz, and random extended chains have $3J_{NH\text{-}\alpha H}$ between 6 and 8 Hz ***(24,34)***. The three-bond coupling constants can also be converted to a dihedral angle using the empirically derived Karplus equation and then used as a torsion angle constraint in molecular dynamics calculations.

5.1.4. NOEs

Amphipathic helical peptides have characteristic NOE patterns that are observed in all helical secondary structures. These include the presence of numerous intense $d_{NN(i,\ i+1)}$ crosspeaks and a large number of weaker interresidue connectivities, such as $d_{\alpha N(i,\ i+1)}$, $d_{\alpha N(i,\ i+3)}$, and $d_{\alpha\beta(i,\ i+3)}$ crosspeaks as is shown schematically in **Fig. 13**. These crosspeaks are observed in the fingerprint region of the NOESY spectrum (**Fig. 8**).

Other techniques, such as circular dichroism, can also provide secondary structure information, albeit overall structure rather than localized. It is highly recommended that a variety of techniques and NMR parameters be used to assess the overall and local secondary structure of an amphipathic peptide. Although the NMR parameters discussed above may not uniquely define a secondary structure when used individually, when taken together they can provide convincing evidence for the qualitative determination of the secondary structure. However, the investigator should always be aware that small linear peptides often sample a variety of conformations in solution and this may complicate the structural interpretation. This has recently been discussed in more detail by Merutka et al. ***(42)***.

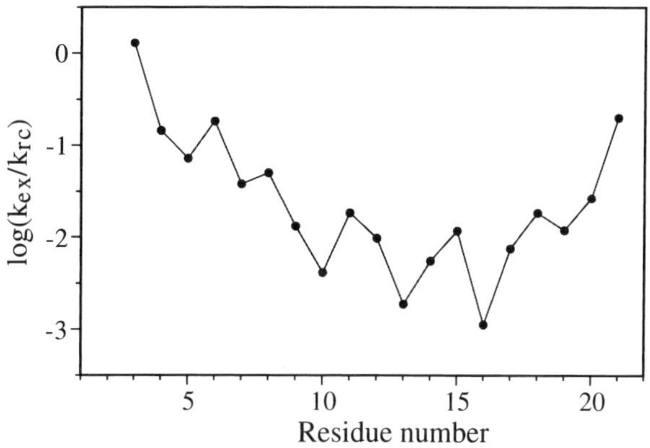

Fig. 12. Relative amide proton exchange for an amphipathic helical peptide, $(KLAKKLA)_3$, plotted as $\log(k_{ex}/k_{rc})$ vs position in the peptide sequence. k_{ex} and k_{rc} are the observed and intrinsic (in random coil) amide proton exchange rate constants, respectively. Where k_{rc} is calculated based on nearest neighbor residue in the sequence, pH* = 3.06 (pH* is the uncorrected meter reading), and 25°C ***(39,40)***.

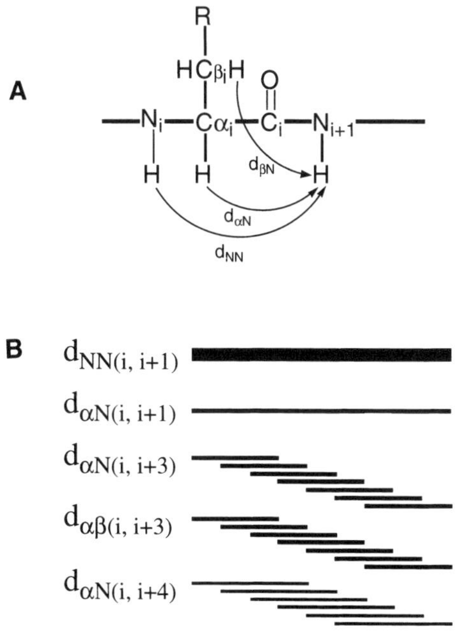

Fig. 13. **(A)** Schematic of the peptide backbone with examples of NOE interactions denoted with arrows. **(B)** Schematic of the typical NOE connectivities observed in an amphipathic helical peptide. The height of the bar is indicative of typical intensity of the NOE interaction.

5.2. Determination of Three-Dimensional Structures

Unlike proteins, in which there can be extensive tertiary interactions determining the 3D fold of the molecule, the structure of amphipathic helical peptides is largely dominated by the secondary structure. The secondary structure can be established via qualitative analysis of NMR data, as discussed previously, or more quantitative analysis as discussed here (keeping in mind the limitations due to possible conformational sampling discussed above). The determination of the 3D structure involves quantitative analysis of NOE crosspeak volumes to obtain distance constraints and quantitative analysis of coupling constants to obtain torsion angle constraints. These experimental constraints are then used in conjunction with calculations to produce a family of structures. There are several software packages available for conducting these calculations, and the specifics for setting up the calculations will depend on the software used. Here we will provide only a brief overview of the steps involved.

The first step of the calculation is to obtain an initial structure, which can be accomplished using several different methods. The most biased method is to start with a known secondary structure, such as an α-helix. Less biased approaches use distance geometry or randomized backbone followed by simulated annealing for the initial structures. After using one or several of these methods to provide an initial structure, further refinement is obtained via restrained molecular dynamics and restrained energy minimization using the NMR-derived distance and torsion angle constraints. Final structures can be used to back calculate a NOESY spectrum for direct comparison with the experimental data and further refinement. The calculations generate a family or families of structures that can be compared on the basis of a root mean square deviation (rmsd).

A careful inspection of particularly the medium or long-range NOEs can reveal if there are multiple conformations in the solution ***(42)***. The recent development of molecular dynamics using time-averaged molecular dynamics may prove useful for molecules with multiple conformations ***(43,44)***.

NMR constraints and calculations such as those described in this section have been used successfully on several amphipathic peptides and have provided information about more long-range structural effects such as bending of the helix axis ***(16,45,46)***.

6. Concluding Remarks

By necessity we have presented only a limited view of the applications of solution proton NMR spectroscopy to studying amphipathic helical peptides. Recent advances in isotopic labeling of proteins have greatly increased the size of the molecule that can be studied as well as increasing the repertoire of experiments that can be used for resonance assignment and structure determination *(2)*. Even for smaller peptides, isotopic labeling can be useful for resolving spectral overlap via multidimensional heteronuclear NMR experiments *(5,47)*. Isotopic labeling can also be used in conjunction with X-filtering to distinguish intermolecular from intramolecular interactions *(48)*. Also, solid-state NMR techniques have been very successful in investigating amphipathic peptides in environments, such as bilayers, which are not accessible to solution techniques *(49)*.

Acknowledgments

The authors would like to thank Karol Maskos for collecting the original data for **Figs. 1**, **7**, **8**, and **10** and for the initial work on the $(KLAKKLA)_3$ peptide. We would also like to thank Dale Treleaven for critical reading of the manuscript. This work was supported by an NSF/EPSCoR grant [(RII/EPSCoR)LEQSF(1992-96)-ADP-01].

References

1. Kessler, H., Gehrke, M., and Griesinger, C. (1988) Two dimensional NMR spectroscopy: background and overview of the experiments. *Angew. Chem., Int. Ed. Engl.* **27**, 490–536.
2. Clore, G. M. and Gronenborn, A. M. (1991) Two-, three-, and four-dimensional NMR methods for obtaining larger and more precise three-dimensional structures of proteins in solution. *Ann. Rev. Biophys. Biophys. Chem.* **20**, 29–63.
3. Bax, A. and Grzesiek, S. (1993) Methodological advances in protein NMR. *Acc. Chem. Res* **26**, 131–138.
4. Redfield, C. (1993) Resonance assignment strategies for small proteins, in *NMR of Macromolecules: A Practical Approach* (Roberts, G. C. K., ed.), IRL, Oxford, pp. 71–100.
5. Markley, J. L. and Kainosho, M. (1993) Stable isotope labelling and resonance assignments in larger proteins, in *NMR of Macromolecules: A Practical Approach* (Roberts, G.C.K., ed.), IRL, Oxford, pp. 101–152.
6. Wüthrich, K. (1995) NMR—this other method for protein and nucleic acid structure determination. *Acta Crystallogr., Sect. D* **D51**, 249–270.
7. Clore, G. M. and Gronenborn, A. M. (1991) Applications of three- and four-dimensional heteronuclear NMR spectroscopy to protein structure determination. *Prog. Nucl. Magn. Reson. Spectrosc.* **23**, 43–92.
8. Dyson, H. J. and Wright, P. E. (1991) Defining solution conformations of small linear peptides. *Ann. Rev. Biophys. Biophys. Chem.* **20**, 519–538.
9. Olson, D. L., Peck, T. L., Webb, A. G., Magin, R. L., and Sweedler, J. V. (1995) High-resolution microcoil ^{1}H-NMR for mass-limited, nanoliter-volume samples. *Science* **270**, 1967–1970.
10. Lee, K. H., Fitton, J. E., and Wüthrich, K. (1987) Nuclear magnetic resonance investigation of the conformation of δ-haemolysin bound to dodecylphosphocholine micelles. *Biochim. Biophys. Acta.* **911**, 144–153.

11. Wishart, D. S., Sykes, B. D., and Richards, F. M. (1992) The chemical shift index: a fast and simple method for the assignment of protein secondary structure through NMR spectroscopy. *Biochemistry* **31**, 1647–1651.
12. Wishart, D. S., Sykes, B. D., and Richards, F. M. (1991) Relationship between nuclear magnetic resonance chemical shift and protein secondary structure. *J. Mol. Biol.* **222**, 311–333.
13. Iwai, H., Nakajima, Y., Natori, S., and Arata, Y. (1993) Solution conformation of an antibacterial peptide, sarcotoxin IA, as determined by ^{1}H-NMR. *Eur. J. Biochem.* **217**, 639–644.
14. Bodon, A., Berthault, P., Segalas, I., Perly, B., and Wróblewski, H. (1995) Solution structure determination by NMR spectroscopy of a synthetic peptide corresponding to a putative amphipathic α-helix of spiralin: resonance assignment, distance geometry and simulated annealing. *Biochim. Biophys. Acta.* **1235**, 169–177.
15. Fan, F. and Mayo, K. H. (1995) Effect of pH on the conformation of backbone dynamics of a 27-residue peptide in trifluoroethanol. *J. Biol. Chem.* **270**, 24,693–24,701.
16. Sipos, D., Andersson, M., and Ehrenberg, A. (1992) The structure of the mammalian antibacterial peptide cecropin P1 in solution, determined by proton-NMR. *Eur. J. Biochem.* **209**, 163–169.
17. Tanford, C. and Reynolds, J. A. (1976) Characterization of membrane proteins in detergent solutions. *Biochim. Biophys. Acta.* **457**, 133–170.
18. Lauterwein, J., Bösch, C., Brown, L. R., and Wüthrich, K. (1979) Physicochemical studies of the protein-lipid interactions in melittin-containing micelles. *Biochim. Biophys. Acta.* **556**, 244–264.
19. Seigneuret, M. and Lévy, D. (1995) A high-resolution ^{1}H NMR approach for structure determination of membrane peptides and proteins in non-deuterated detergent: application to mastoparan X solubilized in n-octylglucoside. *J. Biomol. NMR* **5**, 345–352.
20. McDonnell, P. A. and Opella, S. J. (1993) Effect of detergent concentration on multidimensional solution NMR spectra of membrane proteins in micelles. *J. Magn. Reson. Ser. B* **102**, 120–125.
21. Primrose, W. U. (1993) Sample preparation, in *NMR of Macromolecules: A Practical Approach* (Roberts, G.C.K., ed.), IRL, Oxford, pp. 7–34.
22. Henry, G. D. and Sykes, B. D. (1994) Methods to study membrane protein structure in solution. *Methods Enzymol.* **239**, 515–535.
23. Derome, A. E. (1988) *Modern NMR techniques for chemistry research.* Pergamon Press, Oxford.
24. Wüthrich, K. (1986) *NMR of Proteins and Nucleic Acids.* Wiley, New York.
25. Wishart, D. S., Bigam, C. G., Holm, A., Hodges, R. S., and Sykes, B. D. (1995) ^{1}H, ^{13}C and ^{15}N random coil NMR chemical shifts of the common amino acids. I. Investigation of nearest-neighbor effects. *J. Biomol. NMR* **5**, 67–81.
26. Guéron, M., Plateau, P., and Decorps, M. (1991) Solvent signal suppression in NMR. *Prog. Nucl. Magn. Reson. Spectrosc.* **23**, 135–209.

27. Wider, G., Hosur, R. V., and Wüthrich, K. (1983) Suppression of the Solvent Resonance in 2D NMR Spectra of Proteins in H_20 Solution. *J. Magn. Reson.* **52,** 130–135.
28. Hore, P. J. (1989) Solvent Suppression. *Methods Enzymol.* **176**, 64–77.
29. Piotto, M., Saudek, V., and Sklenar, V. (1992) Gradient-tailored excitation for single quantum NMR spectroscopy of aqueous solutions. *J. Biomol. NMR* **2,** 661–665.
30. Billeter, M., Braun, W., and Wüthrich, K. (1982) Sequential resonance assignments in protein ^{1}H nuclear magnetic resonance spectra. *J. Mol. Biol.* **155**, 321–346.
31. Wüthrich, K., Wider, G., Wagner, G., and Braun, W. (1982) Sequential resonance assignments as a basis for determination of spatial protein structures by high resolution proton nuclear magnetic resonance. *J. Mol. Biol.* **155**, 311–319.
32. Englander, S. W. and Wand, A. J. (1987) Main-chain-directed strategy for the assignment of ^{1}H NMR spectra of proteins. *Biochemistry* **26**, 5953–5958.
33. Celda, B. and Montelione, G. (1993) Total correlation spectroscopy (TOCSY) of proteins using coaddition of spectra recorded with several mixing times. *J. Magn. Reson. Ser. B* **101**, 189–193.
34. Evans, J. N. S. (1995) *Biomolecular NMR Spectroscopy*. Oxford University Press, Oxford.
35. Kuntz, I. D., Kosen, P. A., and Craig, E. C. (1991) Amide chemical shifts in many helices in peptides and proteins are periodic. *J. Am. Chem. Soc.* **113,** 1406–1408.
36. Jimenez, M. A., Blanco, F. J., Rico, M., Santoro, J., Herranz, J., and Nieto, J. L. (1992) Periodic properties of proton conformational shifts in isolated protein helices. *Eur. J. Biochem.* **207,** 39–49.
37. Blanco, F. J., Herranz, J., Gonzales, C., Jimenez, M. A., Rico, M., Santoro, J., and Nieto, J. L. (1992) NMR chemical shifts: a tool to characterize distortions of peptide and protein helices. *J. Am. Chem. Soc.* **114,** 9676–9677.
38. Zhou, N. E., Zhu, B.-Y., Sykes, B. D., and Hodges, R. S. (1992) Relationship between amide proton chemical shifts and hydrogen bonding in amphipathic α-helical peptides. *J. Am. Chem. Soc.* **114,** 4320–4326.
39. Molday, R. S., Englander, S. W., and Kallen, R. G. (1972) Primary structure effects on peptide group hydrogen exchange. *Biochemistry* **11,** 150–158.
40. Bai, Y., Milne, J. S., Mayne, L., and Englander, S. W. (1993) Primary structure effects on peptide group hydrogen exchange. *Proteins* **17,** 75–86.
41. Goodman, E. M. and Kim, P. S. (1991) Periodicity of amide proton exchange rates in a coiled-coil leucine zipper peptide. *Biochemistry* **30,** 11,615–11,620.
42. Merutka, G., Morikis, D., Brüschweiler, R., and Wright, P. E. (1993) NMR Evidence for multiple conformations in a highly helical model peptide. *Biochemistry* **32,** 13,089–13,097.
43. van Gunsteren, W. F., Brunne, R. M., Gros, P., van Schaik, R. C., Schiffer, C. A., and Torda, A. E. (1994) Accounting for molecular mobility in structure determination based on nuclear magnetic resonance spectroscopic and X-ray diffraction data. *Methods Enzymol.* **239,** 619–654.

44. Nanzer, A. P., Poulsen, F. M., van Gunsteren, W. F., and Torda, A. E. (1994) A reassessment of the structure of chymotrypsin inhibitor 2 (CI-2) using time-averaged NMR restraints. *Biochemistry* **33,** 14,503–14,511.
45. Holak, T. A., Engstrom, A., Kraulis, P. J., Lindeberg, G., Bennich, H., Jones. T, A., Gronenborn, A. M., and Clore, G. M. (1988) The solution conformation of the antibacterial peptide cecropin A: a nuclear magnetic resonance and dynamical simulated annealing study. *Biochemistry* **27,** 7620–7629.
46. Bazzo, R., Tappin, M. J., Pastore, A., Harvey, T. S., Carver, J. A., and Campbell, I. D. (1988) The structure of melittin: A ^{1}H-NMR study in methanol. *Eur. J. Biochem.* **173,** 139–146.
47. Fesik, S. W. and Zuiderweg, E. R. P. (1990) Heteronuclear three-dimensional NMR spectroscopy of isotopically labelled biological macromolecules. *Q. Rev. Biophys.* **23,** 97–131.
48. Otting, G. and Wüthrich, K. (1990) Heteronuclear filters in two-dimensional [^{1}H-^{1}H]-NMR spectroscopy: combined use with isotope labeling for studies of macromolecular conformation and intermolecular interactions. *Q. Rev. Biophys.* **23,** 39–96.
49. Opella, S. J., Kim, Y., and McDonnell, P. (1994) Experimental nuclear magnetic resonance studies of membrane proteins. *Methods Enzymol.* **239,** 536–560.

II

Molecular Biology of Antibacterial Peptides

9

Laboratory Production of Antimicrobial Peptides in Native Conformation

Erika V. Valore and Tomas Ganz

1. Introduction

The purification of antimicrobial peptides from natural sources often yields only minute amounts of peptides. Although continuing miniaturization of many analytical methods and bioassays has reduced peptide requirements, X-ray crystallography and 3D NMR still obligate 10 mg of peptide or more, and even higher amounts may be necessary for activity studies in animal models. Chemical peptide synthesis is widely available and frequently yields highly pure bioactive products. However, problems arise with cysteine-rich peptides that must be folded into a specific conformation stabilized by internal disulfide linkages. The natural cellular counterpart of this process involves propeptides that are often much larger than the mature peptide and may contain segments that facilitate folding. It is not surprising that the specific folding of some synthetic cysteine-rich peptides has proven to be such a difficult art. Biosynthesis of recombinant cysteine-rich peptides in bacteria produces misfolded peptides that often form poorly soluble inclusions that must be extracted under reducing and denaturing conditions. Such recombinant peptides must then be refolded. These technical problems are compounded when there is too little natural peptide to serve as a standard for the determination of successful refolding of the (bio)synthetic peptide.

We adapted the baculovirus/insect cell system (**Fig. 1**) to produce cysteine-rich antimicrobial peptides in their native conformation in 1–10 mg quantities *(1)*. The requisite technology is accessible to most laboratories. The general approach consists of introducing a cDNA construct for the desired prepropeptide into the baculovirus genome so that it is transcribed under the

From: *Methods in Molecular Biology, Vol. 78: Antibacterial Peptide Protocols*
Edited by: W. M. Shafer, Humana Press Inc., Totowa, NJ

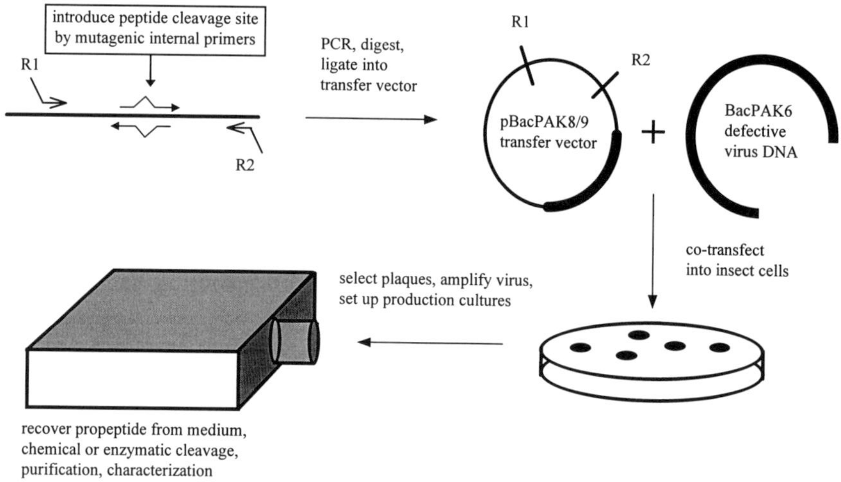

Fig. 1. Outline of the procedure.

control of the very strong polyhedrin promoter, whose normal gene product is nonessential in vitro. The recombinant baculovirus is grown in cultured insect cells. Unlike in their cells of origin, where antimicrobial peptides undergo complete posttranslational processing to mature peptides, the infected insect cells produce the prepropeptide, remove its endoplasmic reticulum-targeting sequence, then, lacking the specific processing enzymes for further peptide maturation, secrete the propeptide into the medium. After several days of culture in serum-free medium, the recombinant propeptide is usually the predominant extracellular protein, and is readily recovered by ultrafiltration. The subsequent conversion of the propeptide to mature peptide can be facilitated by site-specific mutagenesis to engineer specific cleavage sites into the propeptide. The introduction of a methionine residue by site-specific PCR mutagenesis is particularly effective since the modified propeptide can be cleaved chemically with cyanogen bromide. This method has been employed to produce prodefensins and mature defensins HNP-1 and HD-5, and the proprotegrin and mature protegrin PG-3. The method has been essential in the characterization of human defensins from tissues that are in short supply (Paneth cell defensins of the small bowel) and in studies of the structure and functional role(s) of transient antimicrobial peptide precursors. We also find that immunization with recombinant full-length peptides and propeptides produces much higher titer antibodies against the native peptide than immunization with chemically synthesized peptide fragments that may differ in conformation from their native counterparts.

1.1. Preparation of cDNA Fragment for Cloning into the Transfer Vector

A cDNA fragment including the ribosomal binding site, the translation start codon ATG, the entire coding region of prepropeptide and the stop codon is amplified with a pair of PCR primers modified to generate flanking restriction sites. Suitable restriction sites that avoid undesirable cleavage within the cDNA should be chosen to mirror those in the transfer vector cloning region. To minimize spontaneous PCR mutagenesis, high fidelity DNA polymerases are used for amplification. It is advisable to allow at least three base pairs on the "outside" of each palindromic restriction sequence in order to facilitate the digestion of the amplified PCR fragment by the restriction enzymes. The new generation of transfer vectors (e.g., Clontech pBacPAK8/9, Palo Alto, CA) allows the cloning of asymmetric fragments that can greatly simplify the cloning procedure since the fragment orientation within the vector is assured, and self-ligation of the vector is obviated. To generate suitable cleavage sites in the encoded prepropeptide, the cDNA insert can be altered by PCR mutagenesis *(2)*. Because of the simplicity of the process and the limited change in prepropeptide structure, we prefer the substitution of methionine to generate a suitable CNBr-cleavage site. More complex modifications may be necessary if there are methionines present in the desired peptide fragment.

1.2. The Production of a Transfer Vector Construct

The amplified cDNA insert and the transfer vector are digested with the appropriate restriction enzymes, and purified from small cleavage fragments that could inhibit specific ligation *(3,4)*. Suitable purification methods include agarose gel electrophoresis or spin column gel permeation chromatography. The insert is ligated into the vector and transformed into XL-1 Blue strain of *E. coli*. Recombinant clones are picked, grown, and plasmid preparations verified by restriction analysis and DNA sequencing to confirm that there are no undesirable mutations in the cDNA insert sequence. A larger scale plasmid purification of the transfer construct (0.5 mg or more) at high purity is then performed by a commercially available DNA-binding column or ultracentrifugation on a CsCl gradient *(4)*.

1.3. The Production of Recombinant Baculovirus

The transfer construct and Bsu361-digested BacPAK6 (*Autographa californica* nuclear polyhedrosis) viral DNA are cotransfected by lipofectin into insect Sf21 cells. The partially deleted BacPak6 DNA is complemented by sequences in the transfer vector, so that only a homologous recombination event between the transfer vector and the wild-type virus will yield viable virus. Recombinant baculovirus that contains the peptide cDNA is harvested from the culture medium after cotransfection, and individual recombinant clones are isolated as viral plaques. The viral clones are expanded and their supernatants compared for the production of the desired propeptide, e.g., by migration on PAGE, enzyme immunoassay (EIA), Western blot, or functional assay.

1.4. Production of Recombinant Peptide

The following protocol is based on our experience with the production of defensins proHNP-1, HNP-1, proHD-5, and HD-5, and the protegrin precursor proPG3. Sf21 insect cells infected with recombinant baculovirus release propeptide into the culture medium, where it accumulates to a concentration of up to several µg/mL after 60–72 h of infection. For larger scale protein expression *Trichoplusia ni* 5B1–4 (High Five) adherent cells (Invitrogen, San Diego, CA) are used since they grow faster and yield more recombinant protein than Sf21 cells. High Five cells cultured in serum-free medium are infected at exponential growth phase in large cell culture flasks with 10 PFU (plaque forming units) of recombinant virus per cell. Alternatively, large numbers of cells grown in suspension can be infected with baculovirus, but strict temperature control is required for optimal production. At approx 60–72 h after infection, the medium is collected, and cleared of detached cells by centrifugation. Protease inhibitors and/or reducing agents can be added to the medium to minimize oxidation and proteolysis. Prevention of methionine oxidation is critical for chemical cleavage with cyanogen bromide. The medium can be stored at either 4°C for 1–2 wk or can be frozen for longer storage.

1.5. Purification of Recombinant Peptides

Medium containing recombinant peptide is concentrated approx 10- to 20-fold, and diafiltered (for cationic antimicrobial peptides we use 5% acetic acid as a solvent) using a tangential flow concentration apparatus. The concentrate can be lyophilized, and chemically cleaved with cyanogen bromide or a specific enzyme, before the peptide of interest is purified. The purification methodology is specific to each application. For defensins and protegrins, or their precursor peptides, concentrated proteins are separated by a continuous flow preparative gel electrophoresis in a 15.8% acid–urea polyacrylamide gel *(5)*. Fractions containing the peptides are assayed by Coomassie-stained acid-urea or SDS-tricine PAGE *(6)* and further purified by reverse phase HPLC *(7)*.

1.6. Characterization of Recombinant Peptides

Recombinant peptides should be subjected to amino acid sequencing to confirm that the correct peptide has been produced. Their size can be estimated on

SDS-Tricine PAGE *(6)* or determined more accurately by mass spectrometry. To detect conformational alterations that do not change the mass of the peptide, we compare the electrophoretic migration of native and recombinant versions on acid–urea polyacrylamide gels (for cationic proteins) or their retention time on reverse-phase HPLC. In most cases, correct folding is required for bioactivity, and can be confirmed by functional assays, including biochemical, antimicrobial *(7)*, or cytotoxicity assay *(8,9)* depending on the type of recombinant peptide produced.

Recombinant proproteins or recombinant mature proteins conjugated to a carrier such as ovalbumin can also be used as immunogens for the production of polyclonal or monoclonal antibodies. In our experience, some antibodies produced against the propeptide also react well with the mature peptide. With this methodology, we have produced high quality antibodies for immunostaining tissue sections and for immunoprecipitation of native peptides.

2. Materials

2.1. Preparation of cDNA Fragment for Cloning

1. Peptide cDNA template (containing the ribosomal binding site, translation start codon, the entire coding region and stop codon).
2. 25–30 nucleotide oligomer flanking PCR primers (containing a restriction site and three additional complementary nucleotides at the 5' end of each oligomer). It is best to design these using one of many available computer programs that can scan for potentially troublesome sequence patterns. Gel-purify and stock at a concentration of 20 µ*M* in dH_2O.
3. PCR reagents including a high fidelity thermostable DNA polymerase (e.g., *Pfu* DNA polymerase, Stratagene, La Jolla, CA).

2.2. PCR Mutagenesis

1. All reagents as in **Subheading 2.1.**
2. Two internal 25–30 nucleotide oligomer PCR primers with a mutagenic sequence that will replace the codon immediately N-terminal to the mature peptide with a methionine codon (ATG). *See* **Fig. 2B** for an outline of the procedure.

2.3. Preparation of Insert/Transfer Vector for Ligation

2.3.1. Preparation of cDNA Insert/Transfer Vector

1. 5-µg Transfer vector (e.g., pBacPAK1 or 8/9, Clontech, Palo Alto, CA).
2. 2–4 µg cDNA insert.
3. Restriction enzyme (RE).
4. Restriction enzyme buffer.
5. 50:49:1 phenol:chloroform:isoamyl alcohol.
6. Nondenatured ethanol.

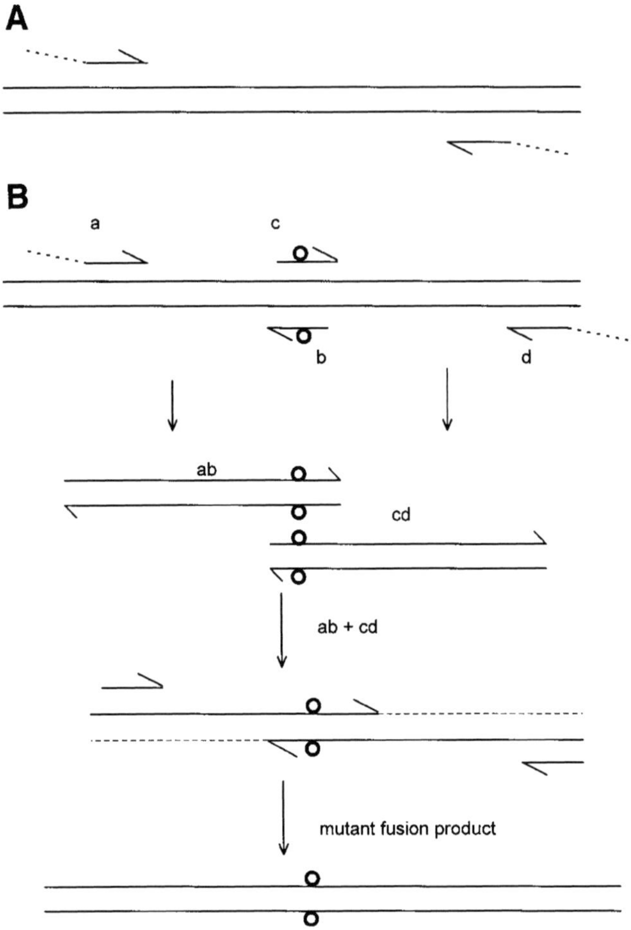

Fig. 2. Construction of HD-5 cDNA insert for recombinant baculovirus by PCR mutagenesis. **(A)** Diagram for the introduction of restriction sites into cDNA. PCR directed mutagenesis is used to introduce restriction sites into cDNA using a cDNA template. Primers are indicated by half arrows and the dotted lines represent the restriction enzyme DNA cleavage sequence. **(B)** Diagram for introduction of restriction sites and exchange of a native codon to a methionine codon. Two sets of PCR reactions are used to generate this construct cDNA as the template. Primers a and d contain restriction enzyme sites (......) and primers b and c are homologous to the cDNA except for a codon (O) that encodes a methionine. Amplification products from each reaction are purified and then used as templates in a subsequent PCR overlap extension (ligation) reaction.

7. 3*M* sodium acetate, pH 5.5.
8. Cow intestine phosphatase (CIP) 1U/µL.
9. 10X CIP reaction buffer.
10. TE pH 7.6: 10 m*M* Tris-HCl, pH 7.6, 1 m*M* EDTA pH 8.0.
11. Low-melting-point agarose (LMP).
12. TBE (10X stock): 89 m*M* Tris base, 89 m*M* boric acid, 20 m*M* EDTA, pH 8.0.

2.3.2. Ligation of Insert and Transfer Vector

1. Transfer vector (treated with RE and CIP).
2. Untreated vector.
3. RE-treated and purified insert.
4. 10 m*M* stock ATP.
5. T4 ligase and 10X ligase reaction buffer.

2.3.3. Transformation into XL-1 Blue E. coli

1. Competent XL-1 Blue *E. coli*.
2. Luria Bertani medium (LB): 10 g Bacto-tryptone (Difco, Detroit, MI), 5 g Bacto-yeast, and 10 g NaCl/L. Adjust to pH 7.5 with NaOH.
3. LB agar plates containing 50 µg/mL ampicillin.

2.3.4. Colony Screen

1. PCR reagents.
2. Primers:
   a. Complementary to the transfer vector.
   b. Complementary to the cDNA insert.

2.3.5. Large-Scale Plasmid Preparation

1. LB medium.
2. Reagents for bacterial lysis.
3. DNA binding column or reagents for CsCI purification.

2.4. Production of Recombinant Baculovirus

1. Insect cell lines (Sf21 or Sf9).
2. Insect culture medium (e.g., Grace's media, Gibco-BRL, Grand Island, NY) containing 10% heat-inactivated FBS and antibiotics.
3. SF-medium: Serum-free insect culture medium without antibiotics (e.g., SF900 media, Gibco-BRL).
4. Bsu361-digested baculovirus DNA (BacPAK6, Clontech).
5. Baculovirus transfer vector containing cDNA insert.
6. Lipofectin (Gibco-BRL).
7. Lux tissue culture plates, 60 mm (Nunc, Roskilde, Denmark).
8. SeaPlaque Agarose (FMC Bioproducts, Rockland, ME).

2.5. Production of Recombinant Peptide

1. Sf21 cells or High Five cells (Invitrogen, San Diego, CA).
2. Serum-free medium: Sf900 medium for SF21 cells (Gibco) or Ex-cell 405 medium for High Five cells (JRH Biosciences, Lenexa, KS).
3. 150-cm^2 culture flasks or suspension culture flasks.
4. Protease inhibitors.
5. 0.1*M* ß-mercaptoethanol in sterile dH_2O (2000X stock, reducing agent).

2.6. Purification of Recombinant Peptide

1. Tangential flow concentrator (Filtron, Northborough, MA) with 10-kDa Omega membrane (Minisette, Filtron).
2. Diafiltration medium; 5% acetic acid (HOAc), 0.05 m*M* ß-mercaptoethanol (ß-ME).
3. Cyanogen bromide solution (1*M* CNBr dissolved in 6*M* guanidine HCl, 0.2*M* HCl made immediately before use).
4. Dialysis apparatus.
5. Continuous flow electrophoresis apparatus (Model 491 Prep Cell, Bio-Rad, Hercules, CA).
6. Reagents to make a 15.8% acid-urea gel *(5)*: urea; glacial acetic acid; solution A: 60% acrylamide, 1.6% *bis*-acrylamide; TEMED (*N,N,N',N'*-Tetramethylethylenediamine); 10% ammonium persulfate (APS); AG501-X8 mixed bed ion exchange resin (Bio-Rad, Hercules, CA); loading buffer: 9*M* urea (deionized) in 5% acetic acid, methyl green dye.
7. Transfer buffer: 5% HOAc.
8. Elution buffer: 5% HOAc, 0.05 m*M* ß-ME.
9. RP-HPLC with a 4.6 × 250 mm Vydac C18 column (Separations Group, Hesperia, CA).
10. Trifluoracetic acid.
11. HPLC quality water and acetonitrile.

3. Methods

3.1. Preparation of cDNA Fragment for Cloning

3.1.1. PCR Mutagenesis: Insertion of Restriction Enzyme Sites

1. Prepare 5–6 tubes of each reaction: 3 ng cDNA template, 200 μ*M* each dNTP, 1.0 μ*M* each primer, 1X reaction buffer, and 2.5U/assay *Pfu* polymerase.
2. Complete 35 cycles of PCR as follows: 94°C for 1 min, 55°C for 1.5 min, 72°C for 2 min.
3. Analyze 5 μL PCR product by separating on a 1–2% agarose gel in TBE.
4. Pool and ethanol precipitate PCR products; extract sequentially with equal volumes of phenol, and chloroform. Ethanol precipitate by adding 1/10 vol of 3*M* sodium acetate pH 5.5 and 2 vol nondenatured ethanol. Incubate overnight at –20°C.

5. Separate PCR-amplified cDNA on a preparative low melting point agarose gel, cut out correctly sized bands, and extract cDNA from the agarose. There are several techniques that are suitable. Electroelution, or agarase digestion, or a commercial kit such as Bio-Rad Prep-A-Gene can be used to isolate pure cDNA.

3.1.2. PCR Mutagenesis: Insertion of Restriction Enzyme and CNBr Cleavage Sites

1. PCR amplification #1: Using a cDNA template, amplify cDNA fragments to incorporate restriction enzymes sites on one end and the methionine mutation of the other end (fragments ab and cd as shown in **Fig. 2B**). PCR parameters are the same as in **step 1** and **2** of **Subheading 3.1.1.** (*see* **Note 1**).
2. Complete **steps 3–5** as in method **Subheading 3.1.1.**
3. PCR amplification #2: Use a mix of 5–10 ng of each fragment (i.e., ab + cd) as the template for this PCR (ligation) overlay extension reaction. Complete 35 cycles of PCR amplification as in **step 1** and **2** of **Subheading 3.1.1.**
4. Complete **steps 3–5** as in **Subheading 3.1.1.**

3.2. Production of Transfer Vector Construct

3.2.1. Preparation of Transfer Vector/cDNA for Ligation

3.2.1.1. Restriction Digest

1. Digest transfer vector and cDNA insert (in separate tubes) with the appropriate restriction enzymes. In a total volume of 100 µL combine approx 50 U restriction enzyme, 5 µg vector or 2–4 µg insert, and 1X reaction buffer. Digest at 37°C for 1–2 h.
2. Use 5 µL of the cleaved material to determine the efficiency of digestion by analysis on a 1% agarose minigel.
3. Extract proteins with an equal volume of phenol/chloroform solution.
4. Ethanol-precipitate overnight at –20°C.

3.2.1.2. Phosphatase Treatment of Transfer Vector

1. Centrifuge DNA at 14,000*g* for 30 min, discard supernatant, wash pellet with 70% ethanol, and dry 5–10 min at 37°C. If a transfer vector with non-complementary cloning ends (e.g., BacPAK8, 9) is used, skip to **step 4**.
2. Treat the transfer vector with CIP to prevent self ligation; resuspend vector DNA pellet in dH_2O, add 1X CIP reaction buffer, and 1U CIP in a final volume of 100 µL. Incubate at 37°C for 15 min, again add 1U CIP and incubate at 37°C for 15 min.
3. Extract proteins with phenol/chloroform and ethanol-precipitate DNA.
4. Resuspend vector DNA to a concentration of 50 ng/µL in TE pH 7.6.

3.2.1.3. Gel Purification

1a. Resuspend insert DNA pellet in 25 µL TE pH 7.6, separate on a preparative 1–2% low melting point (LMP) agarose gel in TBE. Cut out insert band and extract DNA as in **step 5** of **Subheading 3.1.1.**

1b. Alternative: Purify insert on a commercially available spin column.
2. Check for purity and estimate concentration by subjecting an aliquot of the insert DNA to agarose electrophoresis.

3.2.2. Ligation of Insert and Transfer Vector

1. In the ligation reaction use a ratio of eight molecules insert to one molecule transfer vector. Use approx 100 ng vector and an appropriate amount of insert, 0.5 m*M* ATP, 1X ligation buffer, and 3–5 U T4 ligase in a final volume of 20 µL. Incubate 15°C overnight.
2. Prepare controls as in **step 1**, but with the following changes (*see* **Note 2**):
   a. Vector, treated with restriction enzyme (RE) and CIP, no ligase added.
   b. Vector, treated with RE and CIP; add ligase but no insert.
   c. Vector treated with RE but not CIP; add ligase but no insert.
   d. Uncut vector. Incubate at 15°C overnight.

3.2.3. Transformation into XL-1 Blue E. coli

1. To 200 µL competent XL-1 Blue *E. coli*, add 10 µL ligation mix, or 10 µL control mix.
2. Incubate on ice for 30–60 min.
3. Heat shock bacteria by incubating at 37°C for 5 min, then place on ice for 5 min.
4. Add 1 mL of LB media to each sample and incubate in a shaking water bath set at 37°C for 1 h.
5. Centrifuge cells 10 min at 300*g* at room temperature, discard supernatant.
6. Gently resuspend bacteria in 200 µL LB media.
7. Plate 66 µL bacterial suspension (in triplicate) onto agar LB plates containing 50 µL/mL ampicillin.
8. Make two plates of each control.
9. Allow bacterial colonies to grow by incubating overnight at 37°C.

3.2.4. Colony screen

1. Pick 10–20 well-separated bacterial colonies and transfer to individual (4-mL) tubes containing 1 mL LB medium (*see* **Note 3**).
2. Incubate in a shaking water bath for 2 h at 37°C.
3. Bacterial suspensions from each colony are screened by either PCR analysis (one primer complementary to the insert and one primer complementary to the transfer vector) or miniprep plasmid *(3)* restriction digests (use an "uncut vector" colony from **Subheading 3.2.3.** for negative control).
4. The cDNA insert from the colony chosen for further use should be sequenced to verify that no mutations have occurred during the PCR amplification.

3.2.5. Large-Scale Plasmid Preparation

1. Bacteria containing the plasmid with the correct insert are subcultured in 1 L LB and grown overnight at 37°C in a shaking water bath.

2. The bacterial pellet is lysed by standard lysis procedures and plasmid purified either by CsCl gradient or on a DNA binding column (e.g., Qiagen, Chatsworth, CA).

3.3. Production of Recombinant Baculovirus

3.3.1. Transfection and Recombination of Transfer Vector with Baculovirus

1. Seed 1.5 × 10^6 exponentially growing Sf21 cells in each of two, 60-mm LUX tissue culture dishes containing 3 mL culture medium (insect cell medium with serum and antibiotics) gently tilt dish to distribute cells evenly over the surface (*see* **Notes 4** and **5**).
2. Incubate cells overnight in a humidified 27°C incubator.
3. Remove medium from dish and gently wash monolayer with 4 mL SF-media.
4. Add 4 mL SF-medium and incubate 10–30 min at room temperature.
5. Meanwhile prepare Lipofectin-DNA complex: in a sterile polystyrene tube mix (*see* **Note 6**): 80 µL dH_2O, 10 µL of 100 ng/µL plasmid DNA, and 10 µL Bsu361 digested BacPAK6 viral DNA (Clonetech).
6. Add 11 µL Lipofectin (1 mg/mL stock) to 99 µL sterile dH_2O in a polystyrene tube and add 100 µL of this mixture to the DNA solution in step 5. Mix gently and incubate at room temperature for 15 min to allow Lipofectin-DNA complexes to form.
7. Replace media in the culture dishes with 3 mL SF-media and add Lipofectin-DNA complex dropwise to the medium while gently swirling/tilting the dish.
8. Incubate at 27°C for 5 h.
9. Add 3 mL medium containing FBS and antibiotics to each dish.
10. Incubate for 60–72 h.
11. Transfer the supernatant containing the recombinant virus to a sterile tube, clarify by centrifugation, and store at 4°C.

3.3.2. Isolation of Recombinant Clones (Plaque Assay)

1. Seed 2–3 × 10^6 exponentially growing Sf21 cells to each of 14 LUX tissue culture dishes (60 mm) containing 3 mL culture medium (insect cell medium with serum) gently tilt dish to distribute cells evenly over the surface of the dish and incubate overnight at 27°C.
2. Dilute cotransfection supernatant (recombinant virus) to: 10^{-1}, 10^{-2}, 10^{-3}, and the wild-type baculovirus to 10^{-5} in culture media.
3. Use four culture dishes for each dilution of recombinant virus, one dish for wild-type baculovirus (positive control) and one dish of diluent alone (negative control). Aspirate medium from each dish, and infect by adding 200 µL viral dilution dropwise to the center of each dish.
4. Incubate 1 h at room temperature.
5. Meanwhile melt 50 mL of 2% SeaPlaque agarose in dH_2O and add an equal volume of prewarmed 37°C culture medium. Aliquot into four tubes and place in 37°C water bath.

6. Aspirate viral solution from each dish then carefully overlay the cell monolayer with 4 mL of the 1% SeaPlaque solution at 37°C (pipet along one side of the plate to avoid disturbing the monolayer). Allow the agarose to harden and overlay with 4 mL cold culture medium (*see* **Notes 7** and **8**).
7. Incubate for 4–5 d at 27°C.
9. Stain cells with Neutral Red by adding 2 mL 0.03% dye in PBS to each dish and incubating at 27°C for 2–3 h.
10. Aspirate off the stain and place the dish upside down in the dark overnight to allow time for the stain to clear from the viral plaques.
11. Pick well-separated individual viral plaques by punching out the agarose with a sterile Pasteur pipet and transfer to tubes containing 500 μL media (with FCS). Elute virus from agarose plugs overnight at 4°C.
12. Seed 3×10^6 exponentially growing Sf21 cells to LUX tissue culture dishes (60 mm) containing 3 mL culture medium (with FCS); gently tilt dish to distribute cells evenly over the surface of the dish and incubate overnight at 27°C (*see* **Note 9**).
13. Aspirate medium and infect cells by adding 200 μL of the viral plaque eluant to the monolayer. Incubate for 1 h at room temperature, add 6 mL fresh medium, and incubate for 3 d at 27°C.
14. Collect supernatant, clarify by centrifugation at 200*g* for 10 min, and store 4°C in a sterile tube. This is the Passage One viral stock.
15. Analyze the culture supernatants of each viral clone for presence of the desired protein.
16. Store 1–2 mL of the Passage One stock at –70°C. Save an aliquot for further use (*see* **Note 10**).

3.3.3. Preparation of Baculovirus Clones/Virus Stocks

1. Dilute the Passage One stock 10^{-4}, 10^{-5}, 10^{-6}.
2. Infect three plates of Sf21 cells for each dilution as in **Subheading 3.3.2.** (plaque assay) using 200 μL viral dilution per plate. Also include 1–2 plates of positive and negative controls.
3. After visualization of the viral plaques, count the plaques and calculate the viral titer of the Passage One stock.
   The titer (plaque forming units: PFU/mL) of virus stock is:
   Average number of plaques per dish × (1 mL/0.2 mL) × (dilution factor)$^{-1}$ = PFU/mL.
4. Make Passage Two virus stock: Sf21 cells are grown in suspension or in a monolayer. **Note:** in either case, the temperature must be maintained at 27°C, and cells should be in exponential growth phase before infection.
   a. For cells in suspension: Seed 2×10^7 cells in 100 mL media-FBS. Incubate while stirring at 27°C until the cell density reaches $4–5 \times 10^5$ cells/mL (approx 2 d) infect by adding 0.1 PFU/cell.
   b. For cells in monolayer: Seed 5×10^6 cells in each of five 75-cm^2 LUX culture flasks. Incubate overnight at 27°C (cell layer should be 70–80% confluent).

Infect with 0.1 PFU per cell of Passage One virus diluted in 5 mL SF-media. Allow virus to adhere by incubating for 1 h at room temperature then add 15 mL fresh media.

5. Incubate at 27°C for 4–6 d.
6. Clarify supernatant by centrifugation at 200*g* and transfer supernatant to a sterile tube. Freeze 5 mL aliquots at –70°C; this is Passage Two virus stock.
7. Determine the titer of the Passage Two stock using the plaque assay as above.
8. Use the passage two stock to make a large volume (500 mL) of working virus stock (Passage Three stock) and determine the titer as above.

3.4. Production of Recombinant Peptide

3.4.1. Cell Monolayer Infection

1. Seed 8×10^6 cells insect cells in each of 5–10 T150 culture flasks and grow overnight at 27°C in serum-free medium (SF-medium).
2. Infect with 10 PFU/cell (the cells in the flask should be approx 70–80% confluent): Aspirate off medium and replace with 5–10 mL SF-medium containing the appropriate amount of passage three baculovirus stock.
3. Incubate for 1h while rocking very slowly to allow virus to adhere to cells.
4. Add 30 mL SF-medium and incubate 60–72 h at 27°C.
5. Pool media and clarify by centrifugation. For protection of the recombinant peptide, add protease inhibitors and a reducing agent and store 4°C 1–2 wk, or at –20°C for longer storage (*see* **Notes 11–14**).

3.4.2. Infection of Cell Suspension

1. Seed 1×10^8 cells in 250 mL SF-medium. Incubate in stirred culture flask at 27°C until the cell density reaches 1×10^6 cells/mL (approx 2 d).
2. Centrifuge cells at 200*g* for 10 min, discard supernatant, and resuspend cells in 1 L SF-medium.
3. Culture cells while stirring at 27°C until cell density reaches 1.5×10^6 cells/mL (approx 1–2 d).
4. Check viability (should be >80%) and infect cells with 1–5 PFU/cell.
5. Check aliquots at different time points to determine optimal infection time.
6. After infection, centrifuge to clarify culture medium, add protease inhibitors/ reducing agent, and store at 4°C for 1–2 wk or at –20°C for longer storage (*see* **Notes 15–17**).

3.5. Purification of Recombinant Peptides

3.5.1. Concentration and Diafiltration

1. Approximately 1–2 L of medium containing recombinant peptide is concentrated using a Filtron "Minisette" tangential flow concentrator with a 10-kDa mol-wt cutoff Omega membrane to approx 100 mL (*see* **Notes 18** and **19**).

2. Diafilter concentrate with 10 aliquots of 100 mL 5% HOAc, 0.05 m*M* ß-ME (i.e., concentrate, add 100 mL 5% HOAc, concentrate again, and so on). Final sample volume should be around 100 mL.
3. Lyophilize concentrate overnight.

3.5.2. CNBr Cleavage

This reaction should be done in a chemical fume hood.

3.5.2.1. Small Scale CNBr (Analytical)

1. Lyophilize 2 µg purified peptide and resuspend in 25 µL freshly made 1*M* CNBr solution.
2. Overlay with N_2 gas and incubate overnight at room temperature in the dark.
3. Hydrolyze unreacted CNBr by adding 100 µL H_2O and incubating for 1–2 h.
4. Microdialyze vs 2 L 5% HOAc with 1–2 buffer changes.
5. Dry sample and resuspend in sample loading buffer, check cleavage by analysis on an acrylamide gel (*see* **Notes 20–22**).

3.5.2.2. Large-Scale CNBr (Preparative)

1. Resuspend dried sample in 16 mL of guanidine solution, transfer to a 50-mL tube with a plug-seal cap containing 1.6 g CNBr crystal, overlay with N_2 gas, and cap tightly. Mix until most of the crystals are dissolved, cover with foil, incubate overnight at room temperature.
2. Hydrolyze excess CNBr by bringing volume to 50 mL with dH_2O and incubating for 2–3 h at room temperature.
3. Dialyze against 4 L of 5% HOAc for 6–8 h with three changes of dialysate.
4. Lyophilize and prepare for purification by continuous flow electrophoresis.

3.5.3. Continuous Flow Electrophoresis

1. Pour a 15.8% acid-urea preparative gel to 12 cm in a large casting tube (Bio-Rad): Dissolve 43.5 g urea in 64.5 mL water and deionize with AG-501-X8 resin. Filter and add 36 mL solution A, degas, add 7.8 mL TEMED and 2.7 mL 10% APS. Overlay with *n*-butanol saturated water and polymerize overnight.
2. Prerun for approx 6 h at 80 mA constant current towards the negative electrode using 5% HOAc as the running buffer (*see* **Note 23**).
3. Load sample (4–8 mL) and electrophorese at 40 mA (constant current). Collect 5–6 mL fractions in 5% HOAc containing 0.05 m*M* ß-ME.
4. Track protein elution using an inline UV monitor with a chart recorder.
5. Analyze 50–100 µL of each fraction to identify the fractions containing the desired peptide on a Coomassie Blue stained acid-urea or SDS polyacrylamide mini-gel.
6. Lyophilize the chosen fractions, resuspend and pool in 5% acetic acid.
7. Filter out particulates through a cellulose acetate membrane (low protein binding).
8. Store sample at 4°C or –20°C until the next purification step.

3.5.4. HPLC

1. Separate peptide solution on a 4.6 × 250-mm Vydac C18 column (Separations Group), using a 1% acetonitrile increment per minute in 0.1% trifluoroacetic acid (TFA).
2. Lyophilize peptide fractions and resuspend in desired buffer.

4. Notes

1. Only one of the primers needs to contain the codon alteration. The mutant codon should be flanked by at least eight complementary nucleotides on both sides. Annealing temperatures may need to be altered for optimal amplification.
2. Controls:
   a. Negative control: vector, treated with restriction enzyme (RE) and CIP, no ligase added, tests for completeness of cutting.
   b. Negative control: vector, treated with RE and CIP, add ligase but no insert, tests for the effectiveness of the CIP treatment.
   c. Positive control: vector, treated with RE but not CIP, add ligase but not insert, tests for ligase activity.
   d. Positive control: uncut vector, tests the efficiency of transformation.
3. Also grow one or two "uncut vector" colonies in LB medium and save for negative control in colony screen.
4. Sf21 cells grow especially well on the LUX culture plate surface.
5. When plating Sf21 cell monolayers on culture dishes, never swirl the dish as the cells will become more concentrated in the middle. Gently tilt culture dish in all directions immediately after addition of cells.
6. Lipofectin-DNA complexes stick to polypropylene tubes but not to polystyrene tubes.
7. It is important to be sure to aspirate all the viral solution from the dish after infection so that the overlay agarose will adhere to the cell monolayer properly.
8. The agarose overlay is used to prevent the virus from diffusing over the cell monolyer. The virus can then only infect neighboring cells and will form a viral plaque.
9. The initial density of the monolayer is crucial for good plaque formation. If the viral plaques are too small, the monolayer was seeded too densely. Conversely, if the plaques are large and diffuse, the monolayer was seeded with too few cells.
10. Baculovirus is stable at 4°C or below, but it is important that it be protected from light, which can destabilize the virus ***(10)***.
11. It is best to optimize the following parameters for maximum protein production:
    a. The amount of virus (PFU/cell). Check culture supernatants for peptide after infection with 1, 2, 5, 10, 20 PFU/cell.
    b. The time of infection (24–72 h); if allowed to proceed too long the viral infection can lyse cells and proteases will be released resulting in degraded proteins.
    c. The density of the cell monolayer at time of infection.

12. Ex-Cell 405 cells have two advantages for use in recombinant protein production:
    a. They grow faster than Sf21 cells and therefore produce more protein.
    b. They are maintained in serum-free medium and thus do not have to undergo an adaptation period when switching to SF medium for infection.
13. Viral infection is optimal when cells are in exponential growth phase i.e., 70–80% confluent.
14. High multiplicity of infection (m.o.i.) is required for protein production because synchronous infection allows all the cells to produce protein at once. Therefore 10 PFU/cell is used. For production of virus, a low m.o.i. (0.1 PFU/cell) is used.
15. If the methionine is oxidized to methionine sulfoxide the propeptide will not be cleaved by CNBr. The addition of a reducing agent can decrease the oxidation of the methionine residue. At a concentration of 0.05 m*M*, ß-ME will inhibit oxidation of methionine residues but should not affect the disulfide bonds.
16. Strict temperature control is required for infecting cells in suspension.
17. Infection of suspension cultures requires less virus than monolayer infection because there is a higher efficiency of infection.
18. Although the defensin propeptide is smaller than 10 kDa, we found that it was aggregated in the SF-medium. Only small amounts of peptide were lost through the 10-kDa membrane.
19. After testing various techniques, we found that Filtron's tangential flow concentration system worked the best for concentrating the lipid-rich serum-free culture medium. Concentration (1–2 L) and diafiltration (1 L) can be completed in 2–3 h. Thus, minimal degradation of protein occurs during this process.
20. All CNBr steps (except dialysis) should be done in a chemical fume hood. CNBr is very toxic!
21. If the aim is to isolate mature peptide, a large scale CNBr cleavage should be done before the electrophoresis step as the methionine can be oxidized during electrophoresis. However if the final product desired is the proprotein, use the small scale procedure for analytical studies.
22. This method ***(11)*** was optimized for peptides which contain many disulfide bonds.
23. Add 0.1 m*M* sodium thioglycolate to the upper buffer. This will help neutralize reactive oxygen intermediates in the gel.

References

1. Valore, E. V., Martin, E., Harwig, S. S. L., and Ganz, T. (1996) Intramolecular inhibition of human defensin HNP-1 by its propiece. *J. Clin. Invest.* **97,** 1624–1629.
2. Higuchi, R. (1990) Recombinant PCR, in *PCR Protocols.* (Innis, M. A., Gelfand, D. H., Sninsky, J. J., and White, T. J., eds.) Academic, San Diego, pp. 177–183.
3. Ausubel, F. M., Brent, R., Kingston, R. E., Moore, D. D., Seidman, J. G., Smith, J. A., and Struhl, K. (1996) *Current Protocols in Molecular Biology.* Wiley, NY.
4. Maniatis, T., Fritsch, E. F., and Sambrook, J. (1989) *Molecular Cloning.* Cold Spring Harbor Laboratory Press, Cold Spring Harbor, NY, pp. 7–19.

5. Harwig, S. S., Chen, N. P., Park, A. S., and Lehrer, R. I. (1993) Purification of cysteine-rich bioactive peptides from leukocytes by continuous acid-urea-polyacrylamide gel electrophoresis. *Anal. Biochem.* **208,** 382–386.
6. Schagger, H. and von Jagow, G. (1987) Tricine-sodium dodecyl sulfate-polyacrylamide gel electrophoresis for the separation of proteins in the range from 1 to 100 kDa. *Anal. Biochem.* **166,** 368–379.
7. Harwig, S. S., Ganz, T., and Lehrer, R. I. (1994) Neutrophil defensins: purification, characterization, and antimicrobial testing. *Meth. Enzymol.* **236,** 160–172.
8. Lichtenstein, A., Ganz, T., Selsted, M. E., and Lehrer, R. I. (1986) In vitro tumor cell cytolysis mediated by peptide defensins of human and rabbit granulocytes. *Blood* **68,** 1407–1410.
9. Lichtenstein, A. (1991) Mechanism of mammalian cell lysis mediated by peptide defensins. Evidence for an initial alteration of the plasma membrane. *J. Clin. Invest.* **88,** 93–100.
10. Jarvis, D. L. and Garcia, A. J. (1994) Long-term stability of baculoviruses stored under various conditions. *Biotechniques* **16,** 508–513.
11. Villa, S., De Fazio, G., and Canosi, U. (1989) Cyanogen bromide cleavage at methionine residues of polypeptides containing disulfide bonds. *Anal. Biochem.* **177,** 161–164.

10

An Approach Combining Rapid cDNA Amplification and Chemical Synthesis for the Identification of Novel, Cathelicidin-Derived, Antimicrobial Peptides

Alessandro Tossi, Marco Scocchi, Margherita Zanetti, Renato Gennaro, Paola Storici, and Domenico Romeo

1. Introduction

Host defense in mammals involves a cooperative effort by a variety of antimicrobial peptides *(1)*, which are produced by both phagocyte precursors and some epithelia. They are called on to combat invading pathogens at sites of infection and to help protect mucosal surfaces. Although sharing some common properties, such as a high content of basic residues, these peptides vary widely in both size and primary structure.

The neutrophil antimicrobial peptides are synthesized in myeloid precursor cells as prepropeptides, and are stored in cytoplasmic granules as either fully processed products *(2)* or propeptides *(3)*. Molecular cloning of cDNA from bovine, porcine, rabbit, human, mouse, and ovine myeloid mRNA has revealed the presence of a family of antimicrobial peptide precursors, which we have named cathelicidins and which currently counts over 20 members *(4–8)*.

The cathelicidins have highly conserved preproregions, whereas the C-terminal domains, which correspond to the antimicrobial peptides, are highly varied (**Fig. 1**). Some of these peptides had already been isolated as components of the antimicrobial arsenal of neutrophils *(9–13)*, whereas others were deduced from the cDNA of their precursors *(5–8,14–20)* and shown to be antimicrobial after chemical synthesis *(16–19,21)*. The C-terminal peptides in cathelicidins include: (1) Cys-rich peptides, namely the cyclic dodecapeptides

From: *Methods in Molecular Biology, Vol. 78: Antibacterial Peptide Protocols*
Edited by: W. M. Shafer, Humana Press Inc., Totowa, NJ

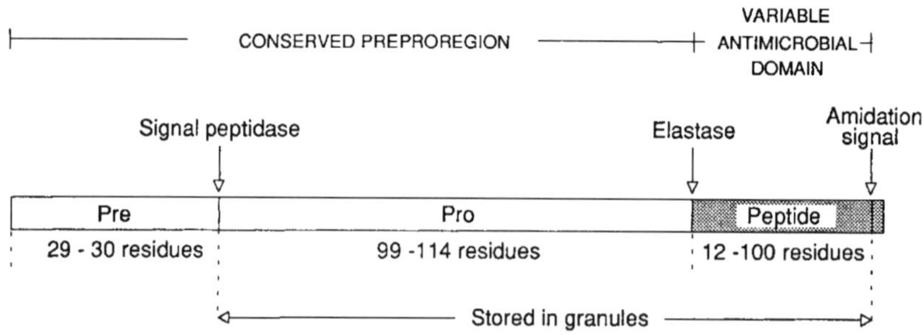

Fig. 1. Schematic representation of a cathelicidin. Sites involved in the processing of the prepropeptide and propeptide are shown.

and protegrins PG-1 to PG-5 ***(5,22–24)***; (2) Pro- and Arg-rich Bac5, Bac7, Bac7.5, prophenin and PR-39 ***(5,15,25–27)***; (3) Trp-rich indolicidin and PMAP-23 ***(17,28)***; and (4) α-helical PMAP-36, PMAP-37, SMAP-29, CAP18(106–142), BMAP-27, BMAP-28, MCLP/CRAMP, and LL-37/hCAP18 ***(5–8,14,16,18–20)***. The precursors of some cathelicidins have been isolated and characterized, and the conserved proregion shown to have an N-terminal pyroglutamic acid and a 1–2, 3–4 disulfide bridging pattern ***(29)***.

Cathelicidin-derived peptides exert a broad spectrum antimicrobial activity at micromolar concentrations, with a wide overlap in specificity but also with significant differences in specific potencies among each other ***(4,6–13, 16–19,21)***. Although several approaches have been used to investigate their mode of action, their killing mechanism(s) is still incompletely understood. Some have been shown to rapidly permeabilize the membranes of susceptible bacteria ***(7,16–18,21)***, with leakage of cellular metabolites and marked reduction of macromolecular biosynthesis ***(30)***.

Extending the search for cathelicidins to various animal species might lead to an expansion of the family, thereby allowing for species correlations. We have thus devised a strategy that permits the discovery of new cathelicidins and a better characterization of the defense system in different animal species. This strategy consists of identification of novel cathelicidins as deduced from their cDNA, identification of their putative antimicrobial domain, and chemical synthesis and biological characterization of the corresponding peptides. A similar strategy can be applied to other precursors of antimicrobial peptides with a conserved domain ***(31)***.

The molecular biological approach for the amplification of novel cathelicidin cDNA sequences is based on the high conservation of their preproregions at the nucleotide level. This strategy is a modified RACE (rapid amplification of cDNA ends) cloning protocol. RACE is a PCR-based technique that was initially developed to facilitate cloning of full-length cDNA ends after production of a partial sequence from a cDNA library ***(32–34)***. We have modified this protocol by using sets of cathelicidin-specific and nonspecific primers to reverse transcribe the 5' and the 3' ends of myeloid mRNAs containing conserved preproregions, and to amplify the cDNA ends by PCR. By using the appropriate primers, overlapping 5' and 3' ends are generated, from which the full-length cDNA sequence is deduced. This method is not only much faster than screening a cDNA library, but is also faster and simpler than the original RACE ***(32)***.

Once a new member of the cathelicidin family has been identified, the putative antimicrobial activity of the C-terminal, nonconserved region needs to be confirmed. We provide advice for the identification of the antimicrobial domain and for the chemical synthesis of the corresponding peptide, including some suggestions for predicting possible difficulties and overcoming them. Finally, we describe assays for determining antimicrobial and membrane-permeabilizing activity of the synthetic peptides.

2. Materials

2.1. Preparation of Myeloid Cell-Enriched Bone Marrow Cells

1. Phosphate-buffered saline (PBS): 137 m*M* NaCl, 1.7 m*M* KCl, 8.1 m*M* Na_2HPO_4, 1,5 m*M* KH_2PO_4, pH 7.4.
2. Lysis buffer: 155 m*M* NH_4Cl, 10 m*M* $KHCO_3$ and 0.1 m*M* EDTA, pH 7.4 at 4°C.
3. Ficoll-Paque is a product of Pharmacia Biotech (Uppsala, Sweden).

2.2. Reverse Transcription and Amplification of cDNA

1. Reverse transcriptase buffer (buffer 1): 250 m*M* Tris-HCl, 375 m*M* KCl, 15 m*M* $MgCl_2$, pH 8.3.
2. PCR reaction buffer (buffer 2): 100 m*M* Tris-HCl, 500 m*M* KCl, 15 m*M* $MgCl_2$, and 0.1% gelatin (w/v), pH 8.3.
3. Proteinase K reaction buffer (buffer 3): 200 m*M* Tris-HCl, 100 m*M* EDTA, and 10% SDS (w/v), pH 8.0.
4. dNTP mixture: 10 m*M* of each of the four deoxyribonucleoside triphosphates, as obtained from Sigma (St. Louis, MO).
5. Proteinase K (1 mg/mL) can be obtained from Merck (Darmstadt, Germany).
6. Moloney murine leukemia virus (M-MLV) reverse transcriptase (200 U/mL) from BRL (Gaithersburg, MD).
7. Primers 1–4 (*see* **Table 1**), and sequence specific primers, were synthesized using an Applied Biosystem 394 DNA/RNA synthesizer (Foster City, CA).
8. RNAse inhibitor: RNAguard (34 U/μL) can be obtained from Pharmacia Biotech.
9. *Taq* DNA polymerase (2.5 U/μL) is from Perkin Elmer/Cetus (Norwalk, CT).

All other reagents are of molecular biology grade.

Table 1
Oligonucleotide Primer Sequences

Primer	Sequence[a]	Specific for
primer 1	5'TCGGATCCCTCGAGAAGC(T)$_{18}$-3'	poly A tail
primer 2A	5'-GCGAATTCTGTGAGCTTCAGGGTG-3'	proregion
primer 2B	5'-CCGAATTCAGCTACAGGGAGGCCGT-3'	proregion
primer 2C	5'-CCGAATTCAGTGTGACTTCAAGGA-3'	proregion
primer 3	5'-TCGGATCCCTCGAGAAGCTT-3'	primer 1 adaptor sequence
primer 4A	5'-AAGAATTCGGAGACTGGGACCATG-3'	bovine proregion
primer 4B	5'-AAGAATTCGGGCTCACCTGGGCACCATG-3'	porcine proregion

[a]Underlined sequences indicate restriction sites.

2.3. Peptide Synthesis, Purification, and Characterization

1. Fmoc-protected amino acids and 2-(1H-benzotriazole-1-yl-)-1,1,3,3-tetramethyluronium tetrafluoroborate (TBTU) from Novabiochem (Läufelfingen, Switzerland); *N*-hydroxybenzotriazole (HOBt) from Sigma (St. Louis, MO); 1-hydroxy-7-azabenzotriazole (HOAt), *O*-(7-azabenzotriazol-1-yl)-1,1,3, 3-tetramethyluronium hexafluorophosphate (HATU), and PEG-PS resins from Perseptive BioSystems (Framingham, MA).
2. Triisopropylsilane (TIPS), phenol, 1,8-diazabicyclo(5,4,0)undec-7-en (DBU), and acetic anhydride from Aldrich-Chemie (Steinheim, Germany); thanedithiol (EDT, stench), trifluoroacetic acid (TFA, corrosive), piperidine (stench), and *N*-methylmorpholine (NMM) from Fluka Chemie (Buchs, Switzerland).
3. Synthesis grade *N,N*-dimethylformamide (DMF, toxic), *N*-methylpyrrolidone (NMP), dichloromethane (DCM), dimethylsulfoxide (DMSO), and HPLC grade acetonitrile (toxic) from Romil Chemicals (Loughborough, UK).

All other reagents and solvents are of analytical grade.

2.4. Biological activity assays

1. Bacteria dilution buffers: 10 m*M* sodium phosphate, 137 m*M* NaCl, pH 7.4 (buffer 4) in the case of MIC and MBC assays, and 10 m*M* sodium phosphate, 100 m*M* NaCl, pH 7.4 (buffer 5) for the membrane permeabilization assay.

2. Mueller-Hinton medium and Sabouraud liquid medium can be obtained from Difco Laboratories (Detroit, MI); and *o*-nitrophenyl-β-*D*-galactopyranoside (ONPG) from Sigma.
3. U-shaped, 96-well microdilution plates from Sarsted (Nümbrecht, Germany).

3. Methods

3.1. Preparation of Myeloid Cell-Enriched Bone Marrow Cells

1. Collect bone marrow cells from ribs soon after slaughtering.
2. Cut open ribs with a cast cutter and quickly transfer the marrow with a spatula to a tube containing 1 m*M* EDTA in PBS, prewarmed to 37°C.
3. Sediment coarse debris and clots and aspirate lipids with a Pasteur pipet.
4. Pass the cell suspension through a screen of filter cloth to remove small clumps and quickly dilute in PBS containing 10 m*M* glucose.
5. Lyse erythrocytes and erythroid cells with 3 vol lysis buffer for 2–4 min at 4°C, and collect intact cells by centrifugation at 200*g* for 10 min.
6. Resuspend the pellet in PBS containing 0.5 m*M* $MgCl_2$ and 10 m*M* glucose, and remove the mature granulocytes by centrifugation through Ficoll-Paque at 800*g* for 40 min. The layer at the Ficoll-Paque/PBS interface, enriched in immature cells of the myeloid lineage, should be drawn off and used for RNA extraction after washing twice with PBS.

3.2. Amplification of Cathelicidin cDNAs

The method used to obtain cathelicidin cDNAs is schematically illustrated in **Fig. 2**.

3.2.1. 3'-End Amplification of Cathelicidin cDNAs

The cDNA 3'-ends of cathelicidins are obtained by reverse transcription of a myeloid total RNA population. We recommend a single-step extraction method *(35)* for obtaining high-quality, undegraded RNA. In our experience, elimination of nonpolyadenylated species is not required (*see* **Notes 1** and **2**).

The primer used for the annealing step is a 36-base oligonucleotide (primer 1, *see* **Table 1**) with a stretch of 18 dT residues and an adaptor sequence with restriction sites required for the cloning step.

1. Prepare total RNA at a concentration >1 µg/µL and place 1 µL in a 1.5-mL tube.
2. Add 1 µL primer 1 (1 pmol/µL), 4 µL of buffer 1 and 8 µL H_2O.
3. Incubate 5 min at 68°C to denature the RNA.
4. Incubate 15 min at 55°C to anneal the mRNA to the primer.
5. Add 2 µL dithiothreitol (100 m*M*), 2 µL of the 10 m*M* dNTP mixture, 1.5 µL of M-MLV reverse transcriptase and 0.5 µL of RNAguard.
6. Mix gently and incubate for 60 min at 42°C.
7. Heat for 5 min at 95°C to denature the reverse transcriptase.

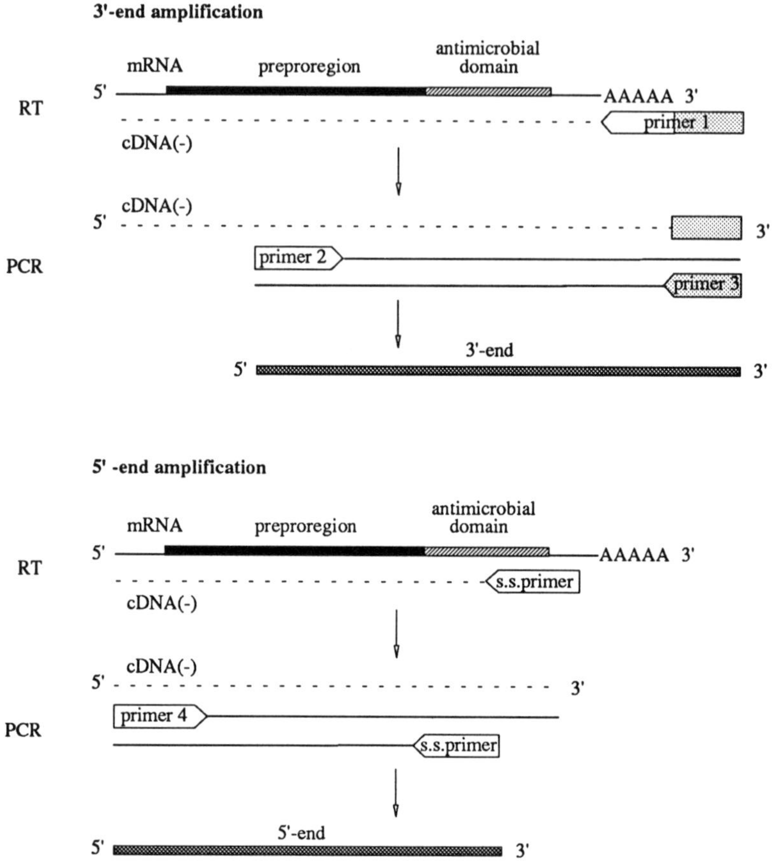

Fig. 2. Schematic representation of the cloning strategy. RT and PCR indicate reverse transcription and PCR amplification steps respectively; s.s. primer indicates sequence specific primer.

The cDNA is then amplified by PCR using a specific primer corresponding to a highly conserved cathelicidin preproregion sequence as sense primer (primers 2A, 2B, or 2C), and the antisense adaptor primer 3. The nucleotide sequences of the three different sense primers used to amplify cathelicidins from different mammals are listed in **Table 1**. All these primers correspond to the most highly conserved sequences in the proregion and include a 5'-end restriction site. Primer 2A is specific for bovine 3' ends. Primers 2B and 2C amplify all the known 3'-ends belonging to the bovine, porcine, and ovine cathelicidins. Primer 2C may also be used to obtain the cDNA sequence of rabbit cathelicidin CAP18. As the number of cathelicidins is continuously growing, new primers may be designed so as to amplify a larger number of cDNAs.

The amplification cycles are carried out on an automated Perkin Elmer Cetus Gene Amp 9600 PCR thermal cycler. Different instruments may require modification of some of the parameters listed below, to optimize the system (*see* **Note 3**):

8. Prepare the following reaction mixture on ice (100 µL final volume): 10 µL of buffer 2, 1 µL of 50 pmol/µL primer 2, 1 µL of 50 pmol/µL primer 3, 1 µL of *Taq* DNA polymerase, 20 µL of cDNA template from step 7, 67 µL H_2O.
9. Carry out 30 cycles of amplification using the following program: 1 min denaturation at 94°C; 1 min annealing at 55°C (*see* **Note 4**), 2 min elongation at 72°C. This is followed by a final elongation step of 5 min at 72°C.
10. Check for the presence of amplified products on 10 µL of the reaction mixture by agarose gel electrophoresis ***(36)***.

It is useful to include a proteinase K digestion step to remove *Taq* polymerase before digesting the PCR products with restriction enzymes, since the polymerase may otherwise remain bound to the DNA and inhibit restriction endonuclease activity ***(37)***.

11. Add 10 µL of buffer 3 and 10 µL of proteinase K to the PCR products.
12. Incubate for 30 min at 37°C and then 10 min at 68°C.

The reaction mixture is then extracted once with phenol-chloroform and once with chloroform before ethanol precipitation ***(36)***. The amplified products are then digested with the appropriate restriction enzymes, the fragments are purified using agarose gel electrophoresis, cloned in the appropriate plasmid vector, and then sequenced according to standard methods ***(36)***.

3.2.2. 5'-End Amplification of Cathelicidin cDNAs

To obtain the 5'-end of cathelicidin cDNAs, reverse transcription specificity is achieved by using a sequence-specific primer derived from a low homology region (mature peptide-coding region and 3'-untranslated region) of the previously determined 3'-end. The experimental conditions used for reverse transcription of the 5'-end are similar to those used for the 3'-end amplification, except that in this case the specific primer is used at a concentration of 0.1 pmol/µL. A low primer concentration increases the specificity.

The 5' cDNA can then be amplified by using primer 4, which corresponds to the conserved 5'-untranslated region of cathelicidins (**Fig. 2**), and a sequence-specific antisense primer that anneals upstream of the primer used for reverse transcription (this "nested" primer increases the specificity and efficiency of the amplification reaction). We have designed two slightly different sense primers 4 based on the highly conserved 5'-untranslated region of cathelicidins: primer 4A amplifies bovine 5' cDNA ends and Primer 4B amplifies porcine 5' cDNA ends (*see* **Note 5**). The PCR products are then processed as in step 11 of **Subheading 3.2.1.**

The full-length cDNA sequence of the identified cathelicidin is then obtained from the overlapping sequences of the 5'- and 3'-end amplification products.

3.3. Synthesis of Antimicrobial Peptides

3.3.1. Sequence Analysis

The cathelicidins that have so far been identified carry a C-terminal domain with antimicrobial activity. This strongly suggests that an antimicrobial peptide should be present in the C-terminal nonconserved region of any new member of this family. However, the mature peptide need not correspond to the entire nonconserved region. Thus, an analysis of the C-terminal region is necessary so as to identify the putative antimicrobial domain, before synthesis is attempted. The following criteria can be of help:

1. A sequence with a high number of positively charged residues; this is very likely to be present in the antimicrobial domain and to be essential for its activity.
2. Similarity to previously characterized antimicrobial peptides derived from cathelicidins. To date, these have fallen into the following broad classes: long peptides (>39 residues), which show an alternating, and sometimes repetitive, distribution of Pro, Arg, and hydrophobic residues; medium-sized peptides (27–39 residues) with an amphipatic α-helical stretch at the N-terminus, and often a hydrophobic C-terminal tail with one or more Pro; short peptides (< 20 residues) with one or two disulfide bridges; and short peptides (13–23 residue) with two or more Trp residues.
3. The presence of a proteolytic processing site after the conserved region (*see* **Fig. 1**). This may help identify the beginning of the antimicrobial peptide, which need not necessarily be at the end of the conserved region ***(15,22)***. Studies with Bac5 and Bac7 indicate that elastase may be the responsible enzyme ***(38)*** so that Val-Xaa or Ala-Xaa sites should be looked for.
4. The prediction of an amphipathic conformation. This is indicated by a sequence of roughly alternating hydrophilic (mainly cationic) and hydrophobic residues. It can easily be identified for α-helical type peptides by the use of a helical wheel projection (*see* **Note 6**).

3.3.2. Synthesis, Cleavage, and Characterization

The synthesis of antimicrobial peptides essentially passes through the following stages:

1. Planning the synthesis, including prediction of difficult stretches, so as to maximize yield (*see* **Note 6**).
2. Carrying out the synthesis, choosing appropriate coupling cycles.
3. Cleavage of the peptide from the resin and deprotection, workup, and purification.

Predicting difficult sequences requires consideration of several factors (e.g., solvation of the polymeric support and growing peptide, inter- and intramolecular interactions in the growing peptide that may lead to steric hindrance, the chemical characteristics of individual Fmoc- and side-chain protected amino acids), and a certain amount of guesswork, as peptide synthesis is a largely unpredictable process. Long peptides, and those that can assume a regular secondary structure during synthesis may lead to problems. The rates and efficiencies of single coupling reactions may depend on the amino acids involved and on the presence of side-chain protecting groups that lead to steric hindrance at the N-terminal amine. These problems and their solution are covered extensively in vol. 35 of this series ***(39)*** and other excellent books ***(40,41)***.

With special regards to the antimicrobial peptides derived from cathelicidins, we have generally found that those belonging to the Pro/Arg-rich class can be synthesized to lengths of over 30 residues quite easily and with good yields. On the other hand, the α-helical peptides often present difficult couplings in the middle of the sequence, leading to collapse of the synthesis and, at best, a formidable purification problem.

Syntheses are carried out on a Milligen 9050 continuous flow instrument, and are normally performed under the following conditions:

1. A 0.05- to 0.1-mmol scale using PEG-PS resins (Fmoc-PAL-PEG-PS for C-terminal amides, preloaded Fmoc-AA-PEG-PS resins for C-terminal carboxylic acids, where AA is the required amino acid).
2. A fourfold excess of Fmoc protected amino acid, HOBt and TBTU (1:1:1) for each coupling, and a 1.7-fold excess of NMM with respect to amino acid.
3. DMF as solvent.
4. 45-min coupling cycles at a flow rate of 3 mL/min.
5. 12-min Fmoc deprotection cycles, using 20% piperidine in DMF, at a flow rate of 3 mL/min.
6. When present, Gln, Asn, and His side chains are protected with the trityl (Trt) group; Lys and Trp side chains with the *t*-butoxycarbonyl (Boc) group; Asp, Glu, Ser, Thr, and Tyr side chains with the *t*-butyl (*t*Bu) group; and Arg side chain with the 2,2,5,7,8-pentamethylchroman-6-sulfonyl (Pmc) group. Cys may be protected with either the Trt or *S*-acetamidomethyl (Acm) groups depending on the method which will eventually be chosen for disulfide bridge formation ***(39)***.

If potentially difficult stretches are predicted (*see* **Note 6**), one or more of the following options may help avoid poor couplings:

7. Increasing the coupling time (e.g., to 60 min).
8. Using a different solvent, such as DMSO/NMP (1:4) or the so called "magic mixture" (*see* **Note 7**).
9. Increasing the Fmoc deprotection time or adding the more powerful base DBU (at 0.7%), to the 20% piperidine cleavage mixture (*see* **Note 8**).
10. Heating to 40–48°C (*see* **Note 9**).
11. Increasing the excess of amino acid and activators.
12. Using different activators, such as the recently available HOAt and HATU instead of HOBt and TBTU.

Any or a combination of **options 7–12** may easily be included in synthesis protocols. **Options 7–9** may be frequently included if there is any doubt as to coupling difficulty. **Option 10** should be used with caution (*see* **Note 9**). **Options 11** and **12** increase the cost of a synthesis; they may be included if a stretch of difficult couplings is predicted, or in a repeat synthesis where the sites of difficult couplings are already known.

Cleavage of peptides from the resin and removal of side-chain protecting groups is carried out using TFA with the addition of appropriate scavengers that remove reactive intermediates. These scavengers are chosen in relation to the types of residues and side-chain protecting groups that are present, as is extensively described in the literature ***(39–41)***. We use 92.5% TFA, 2.5% EDT, 2.5% phenol, and 2.5% water as a default mixture. We add 1–2% TIPS if several Pmc-protected Arg residues are present and 2.5% thioanisole if Met, Trp, or Tyr are present. Generally, 10–20 mL of TFA mixture per gram of resin are adequate. The mixture is added to resin which has been previously extensively washed (methanol followed by ether for PEG-PS resins) and vacuum dried, and the reaction left for at least 2 h. The resin is agitated either by slow N_2 bubbling or placing the reaction vessel on an oscillating tray. The former method has the advantage of removing oxygen, which reduces oxidation of susceptible residues such as Trp. To this end, it is recommended that contact of the reaction mixture with iron-containing objects or dust be avoided. The cleavage mixture is then removed from the resin by vacuum filtration on a syntered funnel, which is washed with a small amount of fresh TFA. Depending on the volume of the mixture, the crude peptide can then be directly precipitated with several volumes of cold ether, isopropyl ether, or *t*-butyl ether (*see* **Note 10**), or the mixture previously reduced in volume by rotary evaporation. Care should be taken not to reduce the volume excessively, as reattachment of protecting groups may occur in some cases. Copious precipitation is normally observed immediately on addition of ether; if in doubt as to complete precipitation, the suspension may be left overnight in a freezer before proceeding further. The suspension is then centrifuged, the ether removed, fresh ether is added, and the pellet broken up and ground or sonicated to remove traces of scavenger. After repeating the process several times, the crude peptide is vacuum dried and ready for purification by RP-HPLC (*see* Chapter 2).

To confirm the correct nature of the purified peptide, several methods can be employed. We have used amino acid analysis, but this frequently gives rise to uncertain results, especially as certain residues (e.g., Arg, Lys, Leu, Pro) are often overrepresented in antimicrobial peptides, and other residues (e.g., Trp) are damaged during hydrolysis. Automated Edman degradation gives more accurate results, but is often inaccessible. We find that mass spectroscopic analysis using electrospray ionization/quadropole (ES-MS) or laser desorption/time-of-flight instruments gives accurate results rapidly and easily. ES-MS is compatible with the eluent from RP-HPLC columns, and indeed the HPLC system can be coupled directly to the mass spectrometer. Apart from confirming the presence of the correct peptide structure, this method can give a good estimate of its purity (*see* **Note 11**).

3.3.3. Examples of Syntheses

3.3.3.1. Synthesis of Bovine Bac7(1–35)

1. Sequence: H-Arg-Arg-Ile-Arg-Pro-Arg-Pro-Pro-Arg-Leu-Pro-Arg-Pro-Arg-Pro-Arg-Pro-Leu-Pro-Phe-Pro-Arg-Pro-Gly-Pro-Arg-Pro-Ile-Pro-Arg-Pro-Leu-Pro-Phe-Pro-OH.
2. Characteristics: despite the length, considered to be an easy synthesis. The proline in position 35 could give rise to diketopiperazine formation ***(40,41)*** so that a fast Fmoc-cleavage cycle (7 mL/min, 4 min) was introduced after coupling of Phe-34.
3. Resin: 0.8 g of Fmoc-Pro-Pepsyn-KA (polyacrylamide/Kieselghur), with a substitution of 0.1 mmol/g (*see* **Note 12**).
4. Synthesis conditions: 0.08-mmol scale, 45-min couplings (except prolines before hydrophobic residues, and residues 1–5, 1 h) with fivefold excess of Fmoc-AA (except residues 1–15, which were in eightfold excess). Solvent DMF.
5. Cleavage conditions: 95% TFA, 2.5% EDT, 2.5% phenol, 2 h.
6. Crude: 0.33 g (~100% yield, may contain scavengers) with purity of 75% (integration of HPLC peaks).
7. Remarks: ES-MS after purification showed one component. The synthesis was satisfactory considering the length of the peptide.

3.3.3.2. Synthesis of Porcine PMAP-23

1. Sequence: H-Arg-Ile-Ile-Asp-Leu-Leu-Trp-Arg-Val-Arg-Arg-Pro-Gln-Lys-Pro-Lys-Phe-Val-Thr-Val-Trp-Val-Arg-OH.

2. Characteristics: predicted difficult couplings between residues Arg-1 to Leu-5 and between residues Lys-16 to Trp-21.
3. Resin: 0.82 g of Fmoc-Arg(Pmc)-PEG-PS with a substitution of 0.11 mmol/g.
4. Synthesis conditions: 0.09-mmol scale, 30-min couplings (except Ile-2, Leu-5, all valines, 45 min) with eightfold excess of Fmoc-AA. Capping (*see* **Note 13**) after coupling of Ile-2, Leu-5, and all valines. Solvent DMF; precoupling column wash with DMSO/NMP (1:4).
5. Cleavage conditions: 90% TFA, 2% each of EDT, phenol, thioanisole, TIPS, and H_2O, 2 h.
6. Crude: 0.16 g (60% yield, may contain scavengers) with purity of about 50%.
7. Remarks: Spectroscopic monitoring of Fmoc cleavage indicates that incomplete couplings may have occurred at Arg-10, Pro-15, and Phe-17. The mass of the purified peptide, as measured by ES-MS (2962 Dalton), is in accordance with theory, and is accompanied by weak shoulders with masses greater by 16 and 32 Dalton, probably owing to oxidation of Trp residues (*see* **Note 11**).

3.3.3.3. Synthesis of Porcine PMAP-36(1–20)

1. Sequence: H-Gly-Arg-Phe-Arg-Arg-Leu-Arg-Lys-Lys-Thr-Arg-Lys-Arg-Leu-Lys-Lys-Ile-Gly-Lys-Val-OH.
2. Characteristics: Peptide corresponding to the predicted amphipathic helical region of PMAP-36.
3. Resin: 0.53 g of Fmoc-Val-PEG-PS with a substitution of 0.19 mmol/g.
4. Synthesis conditions: 0.1-mmol scale, 45-min couplings with fivefold excess of Fmoc-AA, except for residues Arg-2, Arg-5, Arg-13, Lys-16 (sixfold excess) and Lys-19 (eightfold excess).
5. Cleavage conditions: 95 TFA, 2.5% each of phenol and H_2O, 2 h.
6. Crude: 0.2 g (80% yield, may contain scavengers). HPLC indicates two main peaks.
7. Remarks: ES-MS indicates the first HPLC peak corresponds to the correct peptide (60%, 2524.6 Dalton) and the second one to a deletion product missing a glycine (40%, 2468.2). Coupling of the glycine residues in this peptide had not been expected to be difficult. The correct peptide, consisting of only the predicted α-helical portion of PMAP-36, shows a good antimicrobial activity ***(16)***.

3.4. Biological Activity

Once pure synthetic peptides have been obtained, we essentially carry out four types of assays to characterize their biological activity:

1. Minimum inhibitory concentration (MIC) assay to determine if the peptide displays antimicrobial activity and against which microorganisms.
2. Minimum bactericidal concentration (MBC) assay to determine if the peptide's activity is bactericidal or bacteriostatic.
3. Bacterial membrane permeabilization assay to investigate the possible mechanisms of action.
4. Hemolytic activity assay, to test potential activity on animal cells.

3.4.1. MIC determination

Bacterial strains are grown overnight to stationary phase in Mueller-Hinton broth at 37°C. Approximately 0.5–1 mL of stationary phase culture is then added to 50 mL of fresh broth and subcultured with stirring for 2–4 h at 37°C, depending on bacterial growth rate, to obtain midlogarithmic phase bacteria. The bacteria are subsequently sedimented at 1000*g* for 10 min and washed once by centrifugation with buffer 4. The bacteria are then resuspended in buffer 4 and their density is determined by measuring the absorbance at 600 nm and referring to previously determined standards. MICs are determined by a microdilution susceptibility test in sterile 96-well microtitre plates. Microorganisms ($1–2 \times 10^4$ CFU/50 µL) are pipetted into the wells, which contain serial 1:1 dilutions (50 µL in Mueller–Hinton broth) of the peptide to be tested. The highest peptide concentration tested is usually 200–400 µg/mL. The plates are then incubated overnight at 37°C and the MIC corresponds to the lowest peptide concentration that completely inhibits bacterial growth, as determined by visual inspection. Controls lack the peptide but contain TFA at a maximal final concentration of 0.01%, as this may be present after peptide purification by RP-HPLC. The MIC assay of fungal species is carried in the same way except that Sabouraud liquid medium is used throughout, the number of fungi added to the wells is $4–8 \times 10^3$ CFU/50 µL and incubation is for 24–36 h.

3.4.2. MBC Determination

To evaluate the bactericidal activity of the synthetic peptides, aliquots of medium taken from the MIC well and the previous 1 or 2 wells showing no visible bacterial growth at the end of the MIC assay are appropriately diluted in buffer 4, plated on Mueller-Hinton agar, and incubated for 16–18 h at 37°C to allow colony counts. The MBC value is defined as the lowest peptide concentration, causing an at least 99.9% reduction in the number of microorganisms, evaluated as colony forming units, added at the beginning of the MIC determination.

3.4.3. Membrane Permeabilization

The effect of synthetic peptides on the permeability of bacterial inner membrane is evaluated by following the unmasking of cytoplasmic β-galactosidase activity, using the normally inpermeant ONPG substrate in the medium. These experiments are carried out with the lactose permease deficient, β-galactosidase constitutive *E. coli* ML-35 strain. Briefly, logarithmic phase bacteria (about 1×10^7 CFU) are incubated at 37°C in 1 mL of buffer 5 containing 1.5 m*M* ONPG, in a plastic spectrophotometer cuvet. At 0 time, a known amount of the synthetic peptide is added and the rate of *o*-nitrophenol production is recorded at 405 nm. The blank, which is performed in the absence of peptide, normally shows no increase in absorbance. The total β-galactosidase activity is measured with bacteria lysed by ultrasonication, and is carried out both in the presence and absence of peptides to exclude any potential interference of these cationic molecules with enzyme activity. The rate of permeabilization is obtained by determining the slope of the curve, once a steady state has been reached, and comparing it to that obtained after sonication of bacteria. Other methods for investigating the interactions of peptides with membranes are described in Chapter 13.

3.4.4. Hemolytic Activity

Erythrocytes are prepared from freshly drawn blood. This is centrifuged, the erythrocyte pellet is washed three times with cold PBS, and a 20% suspension (v/v) is prepared in PBS. The suspension is brought to 10% by addition of PBS and increasing amounts (up to 200–400 µg/mL) of synthetic peptide, and incubated at 37°C for 15 min in an Eppendorf tube. The reaction is stopped by addition of an equal volume of cold PBS rapidly followed by centrifugation at 10000*g* for 1–2 min. The supernatant is carefully removed, appropriately diluted if neccessary, and the absorbance (A_{pep}) measured at 350 nm. The A_{blank} is evaluated in the absence of additives and 100% hemolysis (A_{tot}) in the presence of 0.2% Triton X-100. Melittin is used as a positive control at concentrations between 1 and 30 µg/mL. The percentage hemolysis is determined as $(A_{pep}–A_{blank})/(A_{tot}–A_{blank}) \times 100$.

4. Notes

1. To minimize the activity of ribonucleases, which may be accidentally introduced from many potential sources in the laboratory, wear gloves and ensure that all solutions, glassware, and plasticware are RNAse free ***(36)***.
2. Owing to the sensitivity of the PCR method, RNA to be used for amplification should be pretreated with DNAse, to avoid unwanted amplification of contaminating genomic DNA.
3. It is critical to avoid contamination from exogenous nucleic acids. Avoid preparing DNA or PCR reagents or carrying out the PCR in a place where large amounts of plasmids, DNA, or PCR products are handled, and always use gloves and sterile plasticware and reagents ***(36,42)***.
4. Sequence specific antisense primers should have a similar length and GC content to primers 2, 3, and 4 (optimal annealing at 55°C). Suggestions for the design of PCR primers are included in volume 15 of this series ***(43)*** and in other manuals ***(36,42)***.
5. This protocol provides a faster and simpler amplification method compared to the original 5' RACE ***(32)***. However, difficulties in amplification of cathelicidins

may be experienced in some cases, using these two primers, owing to mismatches in the sequence. If this is the case, the original method of Frohman ***(32)*** may be used. Briefly, the 3'-end of the first cDNA strand is modified by the addition of a homopolymer tail using terminal deoxynucleotidyl transferase. This strand is then amplified by PCR using a sense primer, which contains a 3' sequence complementary to the homopolymer tail and a 5' cloning adaptor site sequence, and the sequence specific antisense primer.

6. We have found the PC-based program "Peptide Companion" by V. Lebl, G. Lebl, and V. Krchnak (CoshiSoft/PeptiSearch, Tucson, AZ) to be useful in this respect.
7. This consists of a DMF/NMP/DCM (1:1:1) mixture with 2*M* ethylene carbonate and 0.2% Triton-X 100. It can be used in a precoupling column washing cycle, rather than as the solvent used during coupling. This mixture must however not come into contact with the 20% piperidine Fmoc-cleavage mixture, as a precipitate occurs. DMSO should not be used if easily oxidisable residues (Trp, Met) are present in the sequence.
8. This Fmoc-cleavage mixture should not be used after introduction of an Asp residue, as DBU causes aspartimide formation.
9. We use a thermostated column and heat the mixtures used for the precoupling wash (*see* **Note 7**) and Fmoc-cleavage with a recirculating bath. Dehydration of residues such as Gln and Asn may result (*see* **ref. *39***). *Note:* Solvents such as DCM have a low flash point.
10. *t*-Butyl ether is the most efficient. Ethers should be free of peroxides.
11. Peaks with lower molecular weight are normally owing to easily identifiable amino acid deletions, or dehydration especially when Asp and Ser residues are present. Peaks with higher molecular weights can be owing to noncovalent adducts (commonly Na^+, +23; K^+, +39; TFA, +114) or covalent modifications (e.g., oxidation, +16). In the former case, these peaks are eliminated by increasing the orifice voltage, as the more energetic collision with buffer gas strips these ions away, although peptide fragmentation may also occur. Oxidation is commonly a problem if Trp or Met residues are present, and may sometimes occur as an artifact of ionization in the spectrometer.
12. This type of resin was used before PEG-PS resins became available
13. Capping can be carried out with 20% acetic anhydride in DMF. This ensures that, if incomplete coupling occurs, the incorrect peptide is blocked.

Acknowledgments

The financial assistance of the Italian National Research Council (CNR), the Italian Ministry of University and Research, and the Universities of Trieste and Udine is gratefully acknowledged.

References

1. Boman, H. G. (1995) Peptide antibiotics and their role in innate immunity. *Annu. Rev. Immunol.* **13,** 61–92.

2. Ganz, T. (1994) Biosynthesis of defensins and other antimicrobial peptides, in *Antimicrobial Peptides, Ciba Foundation Symposium 186* (Boman, H. G., Marsh, J., and Goode, J. L., eds.), J. Wiley & Sons, Chichester, UK, pp. 77–90.
3. Zanetti, M., Litteri, L., Gennaro, R., Horstmann, H., and Romeo, D. (1990) Bactenecins, defense polypeptides of bovine neutrophils, are generated from precursor molecules stored in the large granules. *J. Cell Biol.* **111,** 1363–1371.
4. Zanetti, M., Gennaro, R., and Romeo, D. (1995) Cathelicidins: a novel protein family with a common proregion and a variable C-terminal antimicrobial domain. *FEBS Lett.* **374,** 1–5.
5. Bagella, L., Scocchi, M., and Zanetti, M. (1995) cDNA sequences of three sheep myeloid cathelicidins. *FEBS Lett.* **376,** 225–228.
6. Mahoney, M. M., Lee A.Y., Brezinski-Caliguri, D. J., and Huttner, K. M. (1995) Molecular analysis of the sheep cathelin family reveals a novel antimicrobial peptide. *FEBS Lett.* **377,** 519–522.
7. Skerlavaj, B., Gennaro, R., Bagella, L., Merluzzi, L., Risso, A., and Zanetti, M. (1996) Biological characterization of two novel cathelicidin-derived peptides and identification of structural requirements for their antimicrobial and cell lytic activities. *J. Biol. Chem.* **271,** 28,375–28,381.
8. Popsueva, A. E., Zinovjeva, M. V., Visser, J. W. M., Zijlmans, M. J. M., Fibbe, W. E., and Belyavsky, A. V. (1996) A novel murine cathelin-like protein expressed in bone marrow *FEBS Lett.* **391,** 5–8.
9. Romeo, D., Skerlavaj, B., Bolognesi, M., and Gennaro, R. (1988) Structure and bactercidal activity of an antibiotic dodecapeptide purified from bovine neutrophils. *J. Biol. Chem.* **263,** 9573–9575.
10. Gennaro, R., Skerlavaj, B., and Romeo, D. (1989) Purification, composition, and activity of two bactenecins, antimicrobial peptides of bovine neutrophils. *Infect. Immun.* **57,** 3142–3146.
11. Agerberth, B., Lee, J-Y., Bergman, T., Carlquist, M., Boman, H. G., Mutt, V., and Jornvall, H. (1991) Amino acid sequence of PR-39. Isolation from pig intestine of a new member of the family of proline-arginine-rich antibacterial peptides. *Eur. J Biochem.* **202,** 849–854.
12. Selsted, M. E., Novotny, M. J., Morris, W. L., Tang, Y-Q., Smith, W., and Cullor, J. S. (1992) Indolicidin, a novel bactericidal tridecapeptide amide from neutrophils. *J. Biol. Chem.* **267,** 4292–4295.
13. Kokryakov, V. N., Harwig, S. S. L., Panyutich, E. A., Shevchenko, A. A., Aleshina, G. M., Shamova, O. V., Korneva, H. A., and Lehrer, R. I. (1993) Protegrins: leukocyte antimicrobial peptides that combine features of corticostatic defensins and tachyplesins. *FEBS Lett.* **327,** 231–236.
14. Larrick, J. W., Morgan, J. G., Palings, I., Hirata, M., and Yen, M. H. (1991) Complementary DNA sequence of rabbit CAP18, a unique lipopolysaccharide binding protein. *Biochem. Biophys. Res. Commun.* **179,** 170–175.
15. Pungercar, J., Strukelj, B., Kopitar, G., Renko, M., Lenarcic, B., Gubensek, F., and Turk, V. (1993) Molecular cloning of a putative homolog of proline/arginine-rich antimicrobial peptides from porcine bone marrow. *FEBS Lett.* **336,** 284–288.

16. Storici, P., Scocchi, M., Tossi, A. Gennaro, R., and Zanetti, M. (1994) Chemical synthesis and biological activity of a novel antimicrobial peptide deduced from a pig myeloid cDNA. *FEBS Lett.* **337,** 303–307.
17. Zanetti, M., Storici, P., Tossi, A., Scocchi, M., and Gennaro, R. (1994) Molecular cloning and chemical synthesis of a novel antimicrobial peptide derived from pig myeloid cells. *J. Biol. Chem.* **269,** 7855–7858.
18. Tossi, A., Scocchi, M., Zanetti, M., Storici, P., and Gennaro, R. (1995) Porcine myeloid antimicrobial peptide, PMAP-37, a novel antimicrobial peptide from pig myeloid cells: cDNA cloning, chemical synthesis and activity. *Eur. J. Biochem.* **228,** 941–946.
19. Agerberth, B., Gunne, H., Odeberg, J., Kogner, P., Boman, H. G., and Gudmundsson, G. H. (1995) FALL-39, a putative human peptide antibiotic, is cysteine-free and expressed in bone marrow and testis. *Proc. Natl. Acad. Sci. (USA)* **92,** 195–199.
20. Larrick, J. W., Hirata, M., Balint, R. F., Lee, J., Zhong, J., and Wright, S. C. (1995) Human CAP18: a novel antimicrobial lipopolysaccharide-binding protein. *Infect. Immun.* **63,** 1291–1297.
21. Tossi, A., Scocchi, M., Skerlavaj, B., and Gennaro, R. (1994) Identification and characterization of a primary antimicrobial domain in CAP18, a lipopolysaccharide binding protein from rabbit leukocytes, *FEBS Lett.* **339,** 108–112.
22. Storici, P., Del Sal, G., Schneider, C., and Zanetti, M. (1992) cDNA sequence analysis of an antibiotic dodecapeptide from neutrophils. *FEBS Lett.* **314,** 187–190.
23. Storici, P. and Zanetti, M. (1993) A novel cDNA sequence encoding a pig antimicrobial peptide with a cathelin-like pro-sequence. *Biochem. Biophys. Res. Commun.* **196,** 1363–1368.
24. Zhao, C., Liu, L., and Lehrer, R. I. (1994) Identification of a new member of the protegrin family by cDNA cloning. *FEBS Lett.* **346,** 285–288.
25. Zanetti, M., Del Sal, G., Storici, P., Schneider, C., and Romeo, D. (1993) The cDNA of the neutrophil antibiotic Bac5 predicts a pro-sequence homologous to a cysteine proteinase inhibitor, that is common to other neutrophil antibiotics. *J. Biol. Chem.* **268,** 522–526.
26. Scocchi, M., Romeo, D., and Zanetti, M. (1994) Molecular cloning of Bac7, a proline- and arginine-rich antimicrobial peptide from bovine neutrophils. *FEBS Lett.* **352,** 197–200.
27. Storici, P. and Zanetti, M. (1993) A cDNA derived from pig bone marrow cells contains a sequence identical to the intestinal antimicrobial peptide PR-39. *Biochem. Biophys. Res. Commun.* **196,** 1058–1065.
28. Del Sal, G., Storici, P., Schneider, C., Romeo, D., and Zanetti, M. (1992) cDNA cloning of the neutrophil bactericidal peptide indolicidin. *Biochem. Biophys. Res. Commun.* **187,** 467–472.
29. Storici, P., Tossi, A., and Romeo, D. (1996) Purification and structural characterization of bovine cathelicidins, precursors of antimicrobial peptides, *Eur. J. Biochem.* **238,** 769–776.

30. Skerlavaj, B., Romeo, D., and Gennaro, R. (1990) Rapid membrane permeabilization and inhibition of vital functions of Gram-negative bacteria by bactenecins. *Infect. Immun.* **58,** 3724–3730.
31. Jones, D. E. and Bevins, C. L. (1992) Paneth cells of the human small intestine express an antimicrobial peptide gene. *J. Biol. Chem.* **267,** 23,216–23,225.
32. Frohman, M. A., Dush, M. K., and Martin, G. L. (1988) Rapid production of full-length cDNAs from rare transcrips: Amplification using a single gene-specific oligonucleotide primer. *Proc. Natl. Acad. Sci. USA* **85,** 8998–9002.
33. Ohara, O., Dorit, R. L., and Gilbert, W. (1989) One-sided polymerase chain reaction: the amplification of cDNA. *Proc. Natl. Acad. Sci. USA* **86,** 5673–5677.
34. Schaefer, B. C. (1995) Revolutions in rapid amplification of cDNA ends: new strategies for polymerase chain reaction cloning of full-length cDNA ends. *Anal. Biochem.* **227,** 255–273.
35. Chomczynski, P. and Sacchi, N. (1987) Single-step method of RNA isolation by acid guanidinium thiocyanate-phenol-chloroform extraction. *Anal. Biochem.* **162,** 156–159.
36. Sambrook, J., Fritsh., E. F., and Maniatis, T. (1989) *Molecular Cloning: A Laboratory Manual* (2nd ed.), Cold Spring Harbor Laboratory, Cold Spring Harbor, NY.
37. Crowe, J. S., Cooper, H. J., Smith, M. A., Sims, M. J., Parker, D., and Gewert, D. (1991) Improved cloning efficienty of polymerase chain reaction (PCR) products after proteinase K digestion. *Nucleic Acids Res.* **19,** 184.
38. Scocchi, M., Skerlavaj, B., Romeo, D., and Gennaro, D. (1992) Proteolytic cleavage by neutrophil elastase converts inactive storage proforms to antimicrobial bactenecins. *Eur. J. Biochem.* **209,** 589–595.
39. Pennington, M. W. and Dunn, B. M., eds. (1994) *Peptide Synthesis Protocols—Methods in Molecular Biology* vol. 35, Humana, Totowa, NJ.
40. Grant, G. A. (1992) *Synthetic Peptides. A User Guide*, W. H. Freeman, New York.
41. Atherton, E. and Sheppard, R. D. (1989) *Solid Phase Synthesis—A Practical Approach.* IRL Oxford University Press, Oxford, UK.
42. Innis, M. A. and Gelfand, D. H. (1990) Optimization of PCRs, in *PCR Protocols: A Guide to Methods and Applications* (Innis, M. A., Gelfand, D. H., Sninsky, J. J., and White, T. J., eds.), Academic, San Diego, CA, pp. 3–12.
43. White, B. A., ed. (1993) *PCR Protocols—Methods in Molecular Biology,* vol. 15, Humana, Totowa, NJ.

11

Molecular Biological Strategies in the Analysis of Antibiotic Peptide Gene Families

The Use Oligonucleotides as Hybridization Probes

Charles L. Bevins and Gill Diamond

1. Introduction

Antibiotic peptides are host defense effector molecules broadly distributed throughout the animal kingdom. Many different families of peptides can be identified based on mature peptide structure or, in some cases, by similar propeptide structure *(1)*. These peptides have been identified in a variety of different cell and tissue types. In several instances, members of a single family of antibiotic peptides are distributed in multiple tissues in a single species. In some species, such as human, more than one family of antibiotic peptide has been identified *(2,3)*. These features afford opportunities to better understand the principles underlying the evolution of this system of host defense.

The precise evolutionary history of the various families remains to be determined. Analysis of the genes encoding the antibiotic peptides has revealed patterns of sequence similarity and genomic structural organization that are consistent with divergent evolution from respective ancestral genes within each of the various peptide families. Whereas sequence conservation can yield important clues to the evolution and regulation of antibiotic peptides, this nucleotide similarity also presents significant challenges for molecular approaches towards the study of these peptides. In some cases, the high degree of sequence identity within these gene families could result in ambiguous data from experiments using standard nucleic acid hybridization conditions. In other cases, the unusual patterns of nucleotide sequence similarity have led to new strategies to identify family members in new tissues or species *(4–8)*.

From: *Methods in Molecular Biology, Vol. 78: Antibacterial Peptide Protocols*
Edited by: W. M. Shafer, Humana Press Inc., Totowa, NJ

A widely used method of detecting the presence of a target RNA or DNA sequences in a complex mixture of nucleic acids is through hybridization using a complementary, radioactively labeled nucleic acid probe. Oligonucleotide probes have proven valuable in discrimination between closely related sequences ***(9)***. Oligonucleotide probes, by virtue of their shorter length, as compared to conventional cDNA and genomic DNA probes, are more sensitive to nucleotide mismatches between probe and target sequences. Therefore, through careful attention to hybridization conditions, one can use oligonucleotide probes to unambiguously differentiate between a target sequence that is perfectly complementary to the probe from one that contains one or more nucleotide mismatches. This approach has been useful in analysis of members of antimicrobial peptide gene families with highly similar nucleotide sequences. This chapter will discuss strategies from our studies of antibiotic peptide genes ***(4,5,7,10–12)***, which we have employed to avoid crosshybridization between closely related sequences and/or to utilize patterns of sequence similarity to identify new antibiotic family members.

2. Materials

2.1. End-Labeling Oligonucleotides to High Specific Activity

1. 0.5*M* Glycine (pH 9.0): Glycine (0.375 g, reagent grade) is dissolved in H_2O (7.5 mL). NaOH (1*M* stock solution, approx 0.65 mL) is added to adjust the pH to 9.0. The solution is then diluted to 10 mL with H_2O. Although this buffer is stable at room temperature, we routinely dispense aliquots (0.5 mL) that are stored at –20°C.
2. 100 m*M* $MgCl_2$: Magnesium chloride crystals ($MgCl_2 \cdot 6H_2O$, 0.203 g, reagent grade, formula wt 203.31) is dissolved in H_2O to a final volume of 10 mL. Aliquots (0.5 mL) are stored at –20°C.
3. 100 m*M* DTT: Dithiothreitol (0.154 g, reagent grade, mol wt 154.25) is dissolved in H_2O to a final volume of 10 mL. Aliquots (0.5 mL) are stored at –20°C.
4. NET: 100 m*M* Sodium chloride, 10 m*M* Tris-HCl, pH 8.0, 1 m*M* EDTA (pH 8.0).
5. [γ-^{32}P]: Adenosine 5′ triphosphate (3000 Ci/mmol, 5 mCi/ml, 1.7 μ*M*, DuPont, Wilmington, DE) should be handled according to standard radiation safety practices to minimize exposure.
6. Oligonucleotides: Deprotected, lyophilized oligonucleotides should be diluted to a concentration of 1 mg/mL in H_2O. This concentrated stock should be stored at –20°C. A 1-mL diluted working stock solution at 1 μ*M* should be made. This concentration is conveniently prepared from the 1 mg/mL stock according to the following relationship: μL of stock solution = (0.33 · n [length of oligonucleotide]), which is diluted to 1 mL in H_2O. Note that, to maximize specific activity, the molar quantity of oligonucleotide (approx 10 pmol) is kept less than the molar amount of ATP (approx 25 pmol).
7. T4 polynucleotide kinase (PNK): This enzyme (10 U/μL) may be obtained from Stratagene, La Jolla, CA.

8. Gel filtration mini columns. We use Push columns (NucTrap™, Stratagene). Many other commercial products are also available for separating labeled probe from free nucleotides.

2.2. Empirical Approach to Appropriate Hybridization Stringency

1. 20X SSC: Dissolve NaCl (1750 g), Na_3citrate·$2H_2O$ (875 g, formula wt 294.1) and citric acid (2.5 g, anhydrous, mol wt 192.1) in a final volume of 10 L.
2. Denhardt's solution ***(13)***: Ficoll 400 (10 g, Sigma), polyvinylpyrrolidone (10 g, Sigma), and bovine serum albumin (10 g, fraction V, Sigma) in H_2O to a final volume of 500 mL. The solution is filter sterilized and stored at –20°C in 50-mL aliquots.
3. 20 mg/mL RNA: Yeast RNA (4 g, type VI, Sigma R6625) is dissolved in 185 mL of 0.55*M* Tris base (conveniently prepared by mixing 180 mL H_2O and 5 mL of 2*M* Tris base). The volume is adjusted to 200 mL by the addition of H_2O. Aliquots of 10–20 mL are stored at –20°C.
4. Formamide: Molecular biology grade formamide may be obtained from Sigma. Caution: Formamide is a teratogen and should be used with appropriate caution.
5. 20% SDS: In the chemical fume hood, dilute 100 g sodium dodecyl sulfate in H_2O to a final volume of 500 mL.
6. 0.5*M* EDTA: Ethylenediaminetetraacetic acid disodium salt (46.53 g, formula wt 372) and sodium hydroxide (5.35 g) are placed in a 250-mL graduated cylinder. Deionized H_2O is added to 230 mL. The mixture is stirred until all salts are dissolved (pH approx 7.75). A 1*M* NaOH solution is added until the pH is 8.0 (approx 10–12 mL). The solution is then diluted to 250 mL with deionized H_2O.
7. 1*M* NaOH: Sodium hydroxide (20g) is dissolved in deionized water to a final volume of 500 mL, and the solution is stored at room temperature in a polypropylene bottle.
8. Sealable plastic pouches: We use polyester Kapak/"Scotchpak," available from Fisher and many other general suppliers.
9. Impulse sealer: A durable impulse sealer may be obtained from National Bag Company (Hudson, OH).
10. Dot blot manifold: A dot-blot apparatus may be obtained from Bethesda (Gaithersburg, MD). Whatman 3MM paper may be obtained from Fisher.
11. Membrane marking pen: Marking pens that show acceptably low levels of pigment bleeding in formamide hybridization solution may be obtained from Schleicher and Schuell (Keene, NH).
12. Charged nylon membrane: We use Hybond N+ (Amersham). Many other similar membranes are readily available.
13. Static water bath. We use a Blue M Magniwhirl bath (Blue Island, IL).
14. Hybridization solution: Stocks (500 mL) are made according to **Table 1**, and are stored at room temperature. Precipitate may form in some of these solutions and will redissolve by warming to 37–42°C. Attention should be paid to avoid removing aliquots if a precipitate exists. We routinely swirl these solutions to aid in examination prior to each use.

Table 1
Composition of Solutions for Oligonucleotide Hybridizations

Component	10%	20%	25%	37.5%	50%
20X SSC	125[a]	125	125	125	125
50X Denhardt's	50	50	50	50	50
Formamide	50	100	125	187	250
20% SDS	25	25	25	25	25
RNA	2.5	2.5	2.5	2.5	2.5
H_20	to 500	to 500	to 500	to 500	to 500

[a]All amounts given in mL.

2.3. Empirical Approach to Blot Washing with Appropriate Stringency

1. 2X SSC/0.1% SDS: We generally make 4 L (400 mL 20X SSC, 20 mL 20% SDS, and water to 4 L), and store at room temperature in eight, 500-mL wide-mouth polypropylene bottles. These bottles are a convenient size for warming in an H_2O bath for high-stringency washes.
2. 0.1X SSC/0.1% SDS: We generally make 4 L (20 mL 20X SSC, 20 mL 20% SDS, and water to 4 L). We also store at room temperature in eight, 500-mL wide-mouth polypropylene bottles, and use a distinctly different colored label to avoid possible mix-ups.
3. Shaking water bath. We use a Precision Model 25.
4. Mercury thermometer.
5. Forceps.
6. Two rectangular polypropylene boxes with lids (8 × 6 × 2 1/2 inches) may be obtained from Cole Palmer (Niles, IL).
7. Plastic wrap.

3. Methods

3.1. End-Labeling Oligonucleotides to High Specific Activity

Gloves and protective clothing should be worn in accordance with radiation safety guidelines, and care should be taken to avoid contamination. Investigators should continually monitor themselves and their work area for radioactivity using a portable monitor. Radioactive waste needs to be disposed in accordance with radiation safety guidelines.

1. The following reagents are added in order to a 1.5-mL microcentrifuge tube:
   a. 10 µL Oligonucleotide working dilution (1µ*M*, approx 10 pmol).
   b. 4 µL Glycine (0.5*M*, pH 9.0).
   c. 4 µL $MgCl_2$ (100 m*M*).
   d. 4 µL DTT (100 m*M*).

e. 15 µL [γ-^{32}P]-ATP (approx 25 pmol, approx 1.65×10^8 DPM).
f. 1.5 µL PNK (10 U/µL).

The first five reagents are added and then thoroughly mixed by vortex action. The tube is briefly centrifuged to force the liquid to the bottom of the tube. An aliquot of PNK enzyme (1.5 µL) is then added, the contents gently vortexed, and the tube centrifuge for 5 s in a microcentrifuge.

2. The reaction tube is then incubated at 37°C for 20 min.
3. The reaction is then diluted with 42 µL of NET and placed on a gel filtration column to remove unincorporated nucleotides. This allows one to estimate the specific activity of the labeled oligonucleotide. We use the Push Column system (NuncTrap™) according to the suppliers protocol. In the past, we have used several other systems to separate free nucleotides, including ion exchange chromatography, and have selected NuncTrap™ out of speed and general convenience. Briefly, these gel cartridges are prewet with 80 µL of NET using an attached syringe to force fluid through the column; the reaction sample is then applied and the eluate is recovered in a 1.5-mL microcentrifuge tube; and an 80-µL aliquot of NET is used to elute the remaining product into the same microcentrifuge tube. The column contains the unincorporated nucleotide and is disposed as dry radioactive waste.
4. A 1-µL aliquot of the eluate (total volume approx 135–160 µL) is removed to determine radioactivity content (*see* **Notes 1–4**).

3.2. Empirical Approach to Appropriate Hybridization Stringency

To establish conditions that will afford both sufficient selectivity and strong hybridization, a series of test hybridization conditions are performed using known target DNA sequences on a single blot. Either cloned phage DNA or plasmid DNA are immobilized on a dot-blot membrane, using multiple aliquots to provide rows that can be cut into strips of identical composition. Circular plasmid DNA should be linearized by digestion with a restriction enzyme that cuts once in the multiple cloning site, but not in the plasmid insert. If both phage and plasmid DNA are used on the same blot, equal molar amounts of each should be applied, which will require about 10–15 times less plasmid DNA in weight than phage DNA. We typically use charged nylon membranes for dot blots, and a protocol for preparation of a dot blot is given here. Protocols for the use of uncharged membranes can be found elsewhere ***(14)***.

1. Cut a piece of charged nylon membrane to the appropriate size for the blotting apparatus to be used. Cut one corner of the membrane for unambiguous orientation. Mark the location of rows and columns using a membrane marking pen. Place the membrane onto the surface of water in a plastic box, and allow the membrane to submerge. Leave in water while assembling the blot apparatus. Cut a piece of 3MM paper to a size appropriate for the manifold; wet in H_2O. Place the 3MM paper in the manifold and then place the nylon membrane on top of it.

Continue to assemble the manifold apparatus according to manufacturer's instructions.

2. Dilute the DNA to a concentration of approx 10 ng/µL for phage DNA or approx 1 ng/µL for plasmid DNA. Prepare 100 µL of DNA solution per strip (e.g., if three hybridization conditions are to be tested, 300 µL of DNA solution is prepared). Add 1/4 vol of 1*M* NaOH (final concentration of DNA solution 0.2*M*) and 1/80 vol of 0.5*M* EDTA (final concentration 5 m*M*). Heat the DNA solution to 65°C for 10 min. Cool to room temperature and centrifuge for 2 s.
3. Place 250 µL H_2O into each well to be used. Place adhesive tape over wells that are not to be used. Apply suction to the manifold device and allow H_2O to be pulled through the well. Apply 125 µL of DNA sample to appropriate wells and allow the sample solution to be pulled through the membrane. Apply 250 µL of 0.2*M* NaOH to each sample well (to wash residual sample to the membrane surface) and allow the wash solution to be pulled through. Dismantle the apparatus, briefly rinse the membrane in water, and then allow the membrane to air-dry. Bake the membrane at 80°C for 2 h (*see* **Note 5**).
4. Cut the membrane into strips that contain the replicate sets of DNA samples (using marks from # 1 as a guide). Label the membranes A, B, C, and so on. Place each membrane strip into a heat-sealable pouch and add 10 mL of hybridization solution. Seal the pouch with a pouch impulse sealer to eliminate pockets of air. Mark the outside of the bag to indicate the concentration of formamide. Incubate all of the bags at 42°C for 15 min (longer is okay but not necessary) (*see* **Note 6**).
5. While the membranes are prehybridizing, add 10 mL of each hybridization solution to a labeled, screw-capped tube. Add 10^7 DPM to each hybridization solution and mix by inversion. When prehybridization is complete, cut a corner of the pouch on a diagonal to create a hole approx 1 cm in diameter. Pour out the prehybridization solution. Using a 10-mL disposable serological pipet, add hybridization solution containing probe to the pouch through the corner hole, keeping bubbles to a minimum. The corner is then double sealed. The outside of the bag is rinsed at the sink, and the hybridization solution is mixed in the bag by dragging the bag gently back and forth across the edge of the sink along the full length of the membrane about a dozen times. All pouches are similarly treated with corresponding hybridization solutions. All hybridizations are then done in a single bath at 42°C overnight (*see* **Notes 7–9**).
6. The next day, the membranes are washed as described in **Subheading 3.3.**

3.3. Empirical Approach to Blot Washing with Appropriate Stringency

1. Place 250 mL of 2X SSC/0.1% SDS in a polypropylene box, and the remaining 250 mL in the polypropylene storage bottle in a shaking H_2O water bath that is set to the appropriate temperature. Use a mercury thermometer to monitor directly the temperature of the wash solution, *inside* the container.

2. A corner of the membrane hybridization pouch is cut and the solution is decanted into a appropriate receptacle for radioactive waste. Three sides of the pouch surrounding the edges of the membrane are then cut, and the membrane is quickly transferred to a plastic tray containing 2XSSC/0.1% SDS at a depth of approx 1 cm. Forceps may aid in the transfer. The lid is placed on the tray, and the tray is gently agitated at room temperature for 10 min.
3. Change the wash solution and repeat the 10-min room temperature incubation with gentle agitation. Repeat this washing step, for a total of six room temperature washes spanning the course of 1 h. These are the low stringency washes. Note that the wash solutions should be treated as radioactive waste, and disposed of accordingly.
4. Transfer the membrane using forceps to the prewarmed wash solution (#1), taking special care to directly check the temperature of this wash solution. Replace the lid on the tray, and gently agitate the tray in the water bath for 10 min. This is the high-stringency wash. Change the wash solution and repeat the 10-min, high-stringency wash once. The total high-stringency wash is 20 min.
5. Remove the final wash solution and rinse membrane with room temperature 2X SSC/0.1% SDS. Blot excess liquid from the membrane quickly with paper towels and immediately cover the membrane in plastic wrap. Do not allow the membrane to dry out. Perform autoradiography.
6. Repeat **steps 1**, **4**, and **5** with each increase in wash stringency. A suggested stepwise increase in stringency for oligonucleotide probes is: 2X SSC/50°C → 2X SSC/57°C → 2X SSC/65°C → 0.1X SSC/50°C → 0.1X SSC/57°C → 0.1X SSC/65°C. One can start the stringency wash at appropriate conditions according to factors discussed in **Subheading 3.4.**, and usually only 3–4 test conditions will achieve desired selectivity (*see* **Notes 10–13**).
7. If the membrane is to be reprobed after an oligonucleotide hybridization, the membrane should be incubated for 20 min at 70°C in 0.1X SSC/0.1% SDS to remove probe. Before using the stripped blot, it should be checked by autoradiography to document effective removal of older probe.

3.4. Hybridization Conditions

To outline a theoretical framework for the specific strategies that our group routinely uses in hybridization experiments, such as the dot-blot experiment described above, as well as Southern and Northern blot experiments, a general discussion of experimental variables is presented. The interested reader is also directed to other more comprehensive reviews of this topic ***(18–20)***.

3.4.1. Overview

Two complementary strands of DNA can form reversible antiparallel complexes, or hybrids. The kinetics of formation and dissociation of complementary nucleotide hybrids are dependent on a collection of experimental variables, including nucleic acid composition and concentration, temperature, and solvent composition. Irreversible attachment of one of the two complementary strands (the target sequence) to a solid support, such as a nylon membrane, allows one to use a second complementary strand as a probe to detect the target sequence in experiments referred to as dot-blot hybridization, Southern blot hybridization, Northern blot hybridization, colony/plaque hybridization, and so on. *(18)*. In practical terms, these experiments involve two steps, hybridization of probe to the target sequences bound to the solid support and subsequent washes to remove nonspecifically associated probe from the solid support. The net effects of the mentioned experimental variables on hybridization and dissociation of hybrids bound to a solid support are similar to those for hybrids free in solution *(17,20–22)*.

For long strands of DNA and RNA, the melting temperature strictly refers to an equilibrium in which 50% of complementary nucleotides have dissociated and 50% remain associated as complementary base pairs. The temperature of complete hybrid dissociation is several degrees higher, but is often loosely referred to as melting temperature also. A distinction will not be drawn in this discussion, since the aim here is to present a general framework of principles to guide in establishing empirically useful experimental conditions, and the interested reader is referred to more detailed discussions of the topic.

3.4.2. Experimental Variables of General Importance in Hybridization Experiments

It is often useful to keep in mind the forces that tend to stabilize nucleic acid hybrids (hydrophobic interactions through base stacking and hydrogen bonding) and the forces that tend to destabilize these hybrids (charge repulsion and entropy). The individual experimental variables can then be altered rationally to optimize a particular experiment. There are five principal variables that are routinely considered in hybridization experiments **(Fig. 1)**:

1. Temperature: Higher temperatures tend to destabilize the hybrid, and the equilibrium shifts to hybrid dissociation (generally referred to as melting).
2. Salt concentration: Higher salt concentrations tend to stabilize the helix *(23,24)*. Salt tends to lessen unfavorable charge interactions and strengthen hydrophobic interactions. At low concentration of salt, the helix melting temperature varies in proportion to the log of monovalent salt concentration, but as the salt concentration approaches 1*M* the effects become much less pronounced.
3. Nucleotide composition: Higher proportions of Gs and Cs, higher GC content, leads to a higher melting temperature of hybrids. The GC content is particularly important for shorter probes, in which the relationship is complex *(25)*. These effects are from a combination of base-stacking interactions and the number of

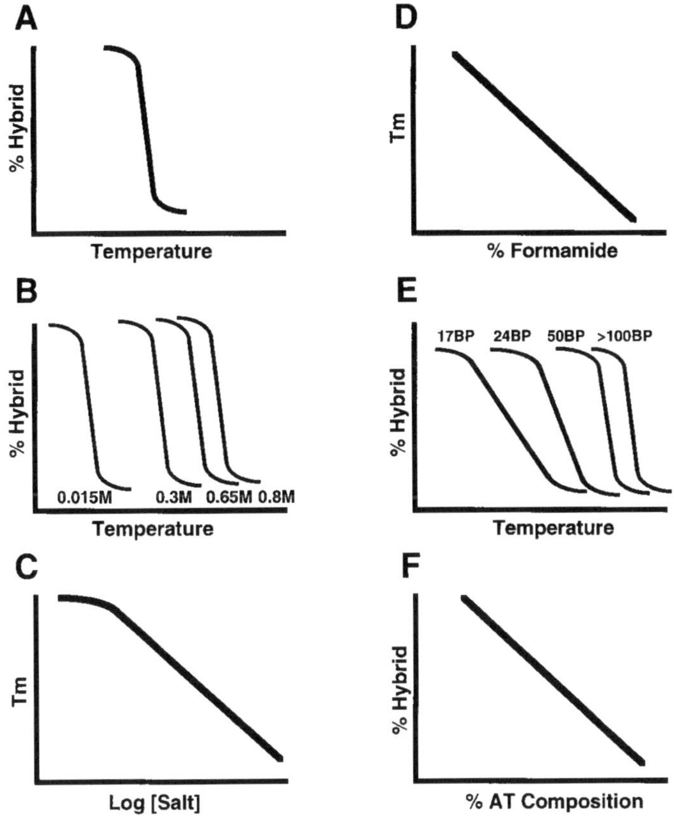

Fig. 1. Diagramatic representation of the effects of experimental variables on hybridization.

hydrogen bonds form between complementary bases in a helix—three hydrogen bonds for GC pairs, and two hydrogen bonds for AT pairs.

4. Formamide concentration: Higher formamide concentration tends to destabilize the helix and result in a lower melting temperature ***(26–28)***. The relationship between melting temperature and formamide concentration is linear, and the melting temperature is lowered by 0.5–0.65°C for each percentage of formamide in the solution.
5. Probe length: Shorter probes melt at lower temperature and manifest less pronounced cooperativity. The relationship is complex, especially with shorter oligonucleotides.

A sigmoid curve is observed when plotting percentage of complementary DNA existing as a hybrid as a function of temperature. This is consistent with cooperative interactions. DNA of infinite length and average GC will melt at approx 100°C in 1*M* salt. Higher salt minimizes unfavorable phosphate interactions along the DNA backbone and tends to strengthen hydrophobic interactions between bases at the core of the helix. Both of these effects of salt stabilize the DNA helix, and increase observed melting temperatures. It is possible to estimate melting temperature (T_m) of relatively long pieces of DNA with average composition, as a logarithmic function of the salt concentration. This relationship holds for monovalent ions, whereas divalent cations have more profound and complex effects and are generally not included in hybridization experiments. One important exception is the buffer for polymerase chain reactions, which generally contains 0.0015*M* Mg^{2+} and 0.05*M* K^+. The cations in this buffer are effectively equivalent to 0.2*M* Na^+ ***(29)***.

$$T_m = 82 + 16.6 \log [Na+] + (0.41 \times GC\% \times 100) - 500/n - (0.65 \times \%\ formamide) \quad (1)$$

Equation 1 shows an empirical formula to estimate the melting temperature (T_m) of DNA hybrids. **Table 2** shows estimated T_m for long DNA of average composition at various salt concentration, along with the relative dilution of the salt solution, SSC, which is commonly used by many laboratories, including ours. As an example of practical information in this Table, for Southern blot washing using SSC if one compares 6X to 2X SSC, the effective T_m is approx 8°C lower in the lower concentration of salt; comparing 2X to 0.1X SSC is equivalent to a change of approx 21°C.

The base composition, and, for relatively short DNA hybrids, nucleotide sequence affects the observed melting temperature of a helix. For relatively long DNA, a linear relationship holds such that the T_m is roughly proportional to 0.41 × %GC × 100 (**Eq. 1**), for GC content between 25 and 75%. This effect is in part due to the extra hydrogen bond that forms between GC base pairs. If one compares the approximate T_m for DNA with GC content from either 25 or 75%, one finds a difference of about 15–20°C. Yet, with shorter DNA hybrids, the effect of GC content is more complex and both composition and sequence are very important ***(25)***.

Formamide is often included in hybridization solutions. It tends to destabilize the helix, resulting in a predictable lowering of melting temperature ***(26–28)***. Generally, 1% formamide in a buffer lowers the T_m of DNA:DNA hybrids by about 0.65°C (**Eq. 1**). This property is useful in adjusting the stringency of hybridization conditions, and allows fine-tuning of the specificity of probe hybridization. In practice, at a constant temperature and salt concentration, a change from 0–50% formamide effectively lowers the melting temperature by approx 25–30°C. Several membrane hybridizations can be independently carried out in a single water bath, with dramatically different

Table 2
Salt Concentration and Approximate Relative Melting Temperatures for Commonly Used Hybridization and Wash Buffers

	$T_m = 102 + 16.6 \log [salt]$	
$[Na^+]$	Dilution of SSC	T_m
0.9*M*	6X	101°C
0.75*M*	5X	100°C
0.3*M*	2X	93°C
0.15*M*	1X	88°C
0.015*M*	0.1X	72°C

T_m, melting temperature.

stringencies, in order to empirically determine the optimal conditions. We have found this approach very helpful with the use of oligonucleotide probes (*see* **Note 9**). Changing salt concentrations to achieve the same spectrum of stringency can dramatically alter the hybridization rates, making comparisons more complex. *Note:* there is somewhat less of an effect on RNA:DNA hybrids (approx 0.45°C for each % formamide *[27]*).

For probes greater than approx 100 nucleotides in length, variation in probe length does not dramatically affect melting temperature. However for shorter probes, length is important and an empirical "fudge" factor has been determined to estimate the effects on melting temperature *(30)*. A rule of thumb that we have found useful is that melting temperature will vary proportionately to $500/n$, in which n is the probe length in nucleotides (**Eq. 1**). Accordingly, the difference between a "long" probe and a 50 mer is approx 8–10°C, and that between a 50 mer and a 24 mer is about 10–12°C. However, shorter probes are much more sensitive to composition and sequence (base stacking energy varies depending on neighboring bases). For oligonucleotide probes we have used **Eq. 1** to give us an estimate of melting temperature for a given probe, and then empirically adjust formamide concentration in the hybridization step, or temperature and salt in the wash step, to select more optimal conditions for a given experiment.

One additional factor that is intimately related to the whole notion of specificity of hybridization experiments is nucleotide mismatches *(31–33)*. Mismatches will destabilize the helix and lead to melting at lower temperature *(34,35)*. An empirical adjustment for the effects of mismatches is a lowering of melting temperature by approx 1.4°C for each percent of mismatch *(31)*. Thus, if a given probe differs in complementary sequence from its target sequence by

15%, the resulting hybrid will likely melt 20°C lower than a perfectly complementary probe.

3.4.3. Kinetics of Hybridization

The generally accepted model for this bimolecular reaction is that a relatively slow nucleation event, whereby the first few nucleotides associate, is then followed by rapid "zipping-up" of hybrid ***(20)***. The nucleation step is influenced by the concentration of nucleic acid molecules and their length, as both of these factors will increase the probability of complementary sequences associating. Empirically, the maximum rate of hybridization is seen at approx 15°C below melting temperature ***(36–38)***. Note that salt has a more pronounced effect on kinetics of hybridization than predicted from its effect on hybrid melting, so we keep salt at 5X SSC (0.75*M*) and vary the temperature and/or formamide concentration to vary the stringency of hybridization.

4. Notes

1. Using this protocol, oligonucleotide probes are end-labeled to a specific activity of approx 10^7–5×10^7 DPM/pmol. This level of specific activity is sufficient to detect a single copy gene in genomic DNA of mammals. Careful attention to avoid using excess oligonucleotide in the end-labeling reaction is important to achieve the desired specific activity.
2. Note that this protocol will efficiently label DNA with a free 5′ hydroxyl group. DNA containing a 5′ phosphate group (e.g., double-stranded DNA product of a restriction enzyme digestion) should be labeled after the phosphate group is removed using a phosphatase enzyme, or should be reacted with polynucleotide kinase under conditions that would promote an exchange reaction ***(15)***.
3. Excess enzyme or excess reaction time can result in reduced labeling efficiency and should be avoided.
4. For quality control regarding the effective use of these NuncTrap™ gel filtration columns, the supplier provides two colored dyes, one of high and the other of low molecular weight, which can visually demonstrate the resolving capabilities of these columns.
5. We generally bake the membrane at 80°C for 2 h, but the membrane manufacturer claims that this is not necessary.
6. Note other investigators have preferred the use of hybridization tubes ***(16)***.
7. We have found hybridization is quite efficient in a static water bath as long as the membrane is uniformly coated with probe-containing hybridization solution by the thorough mixing routine described.
8. We select 24- to 55-base oligonucleotides for most of our hybridization experiments. We select oligonucleotide sequences from either regions of maximal difference (to maximize discrimination between closely related family members) ***(7)***, or from regions of high similarity to identify new family members ***(4)***. For

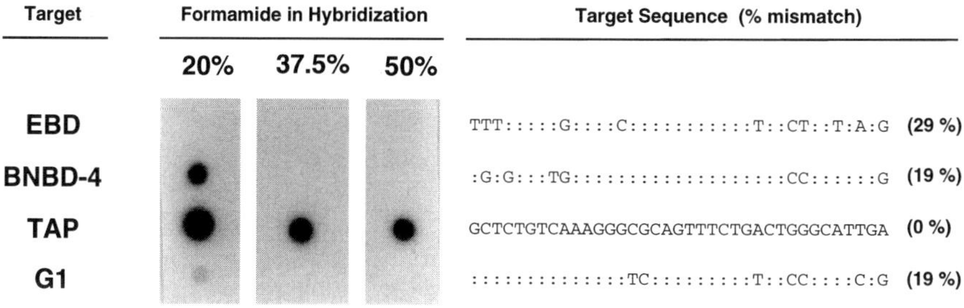

Fig. 2. Hybridization analysis of highly similar β-defensin clones using a 37-base oligonucleotide probe, TAP286a. The phage DNA was spotted on a charged nylon filter for dot-blot hybridization analysis as described in **Subheading 3.2.** Included on the filter was DNA from four β-defensin clones: enteric β-defensin (EBD, *[12]*), neutrophil β-defensin 4 (BNBD-4, unpublished), tracheal antimicrobial peptide (TAP, *[10]*), and G1 (unpublished). Three replicate dot-blot filters were independently hybridized with the oligonucleotide probe under various levels of stringency (20, 37.5, and 50% formamide at 42°C). All three membranes were initially washed at room temperature in 2X SSC, and then at 55°C/2X SSC for 20 min. Autoradiographic exposure time was 16 h (CLB and A. P. Tarver, unpublished). Shown also is a nucleotide alignment of the probe (a perfect match of the TAP sequence) and the other target sequences (:, nucleotide identity, alternative nucleotide indicated). The probe was selected from a region showing relatively low similarity between the various genes.

35- to 40-base oligonucleotides, reasonable conditions for an initial high stringency wash are 2X SSC at 57°C.

9. An example of establishing hybridization conditions to discriminate between highly similar bovine β-defensin cDNAs is shown in **Fig. 2**.
10. When designing an experiment, our experience, which is similar to that reported by others ***(17)***, is that it is better to keep undesirable hybrids from forming (by using higher stringency of hybridization) than it is to attempt to dissociate them through washing the blots. Therefore, we often set conditions of hybridization based on maximal discrimination as outlined above. One exception is if the source of the target sequence is limiting (e.g., RNA from a valuable specimen), where we will use relatively lower hybridization stringency and use progressively higher wash stringency to gain specificity (*see* **Note 13**).
11. Note that the temperature inside the container is not necessarily equivalent to the temperature of the circulating bath.
12. To strip a blot of a cDNA probe, the membrane can be incubated for 20 min at room temperature in 0.2*M* NaOH/1.5*M* NaCl. This protocol should only be used for DNA blots, as RNA is very susceptible to cleavage at high pH. Following this method of probe stripping, the membrane should be rinsed at room temperature in 5% SDS for 20 min to reduce background in subsequent hybridizations.

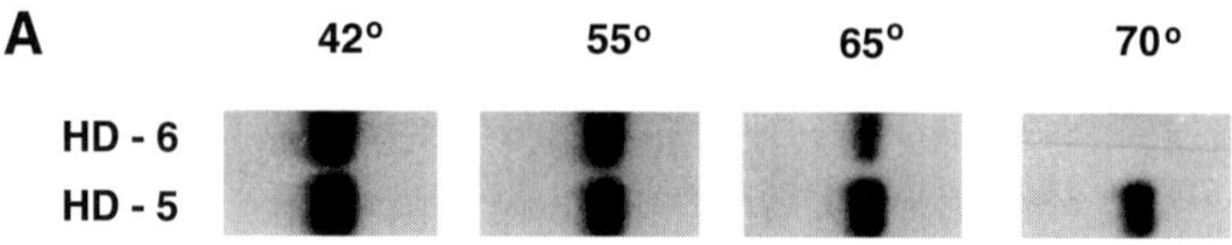


```
B    HD - 6  ::CA:::TG::::::::C::::::::C::::::A:::::::::
     PROBE   CTTGCTGCCATTCTCCTGGTGGCCCTGCAGGCCCAGGCTGAGC
     HD - 5  :::::::::::::::::::::::::::::::::::::::::::
```

Fig. 3. Hybridization analysis of two similar defensin cDNA clones using a 43-base oligonucleotide probe, *D5′-oligo***.** The phage DNA was spotted on a charged nylon membrane as described in **Subheading 3.2.** Included on the filter was DNA from two defensin clones: human defensin 5 (HD-5 *[4]***)**, and human defensin 6 (HD-6 *[5]*). The membranes were hybridized with the oligonucleotide probe in standard hybridization buffer containing 20% formamide (**Table 1**) at 42°C. The filters were initially washed at room temperature in 2X SSC, and then in 2X SSC for 20 min at 42°C. Following autoradiography, the membrane was subsequently rewasched for 20 min in 2XSSC/0.1%SDS at 55°, 65°, and 70°C. (Data reproduced from **ref.** ***4*** with permission.) Shown also is a nucleotide alignment of the probe and the target sequences (:, nucleotide identity, alternative nucleotide indicated).

13. An example of using high stringency wash conditions to discriminate between two similar human defensin cDNAs is shown in **Fig. 3**.

Acknowledgments

Supported by grants AI32738, AI32234, and HL53400 from the National Institutes of Health.

References

1. Boman, H. G. (1995) Peptide antibiotics and their role in innate immunity. *Annu. Rev. Immunol.* **13,** 61–92.
2. Ganz, T. and Lehrer, R. I. (1994) Defensins. *Curr. Opin. Immunol.* **6,** 584–589.
3. Bensch, K. W., Raida, M., Mägert, H.-J., Schulz-Knappe, P., and Forssmann, W.-G. (1995) hBD-1: a novel β-defensin from human plasma. *FEBS Lett.* **368,** 331–335.
4. Jones, D. E. and Bevins, C. L. (1992) Paneth cells of the human small intestine express an antimicrobial peptide gene. *J. Biol. Chem.* **267,** 23,216–23,225.
5. Jones, D. E. and Bevins, C. L. (1993) Defensin-6 mRNA in human Paneth cells: implications for antimicrobial peptides in host defense of the human bowel. *FEBS Lett.* **315,** 187–192.

6. Mahoney, M. M., Lee, A. Y., Brezinski-Caliguri, D. J., and Huttner, K. M. (1995) Molecular analysis of the sheep cathelin family reveals a novel antimicrobial peptide. *FEBS Lett.* **377,** 519–522.
7. Russell, J. P., Diamond, G., Tarver, A., and Bevins, C. L. (1996) Coordinate induction of two antibiotic genes in tracheal epithelial cells exposed to the inflammatory mediators lipopolysaccharide and tumor necrosis factor-α. *Infect. Immun.* **64,** 1565–1568.
8. Tossi, A., Scocchi, M., Zanetti, M., Gennaro, R., Storici, P., and Romeo, D. (this volume) An approach combining rapid cDNA amplification and chemical synthesis for the identification of novel, cathelicidin-derived, antimicrobial peptides in *Current Protocols in Antimicrobial Peptide Research—Methods in Molecular Biology.*
9. Itakura, K., Rossi, J. J., and Wallace, R. B. (1984) Synthesis and use of synthetic oligonucleotides. *Ann. Rev. Biochem.* **53,** 323–356.
10. Diamond, G., Jones, D. E., and Bevins, C. L. (1993) Airway epithelial cells are the site of expression of a mammalian antimicrobial peptide gene. *Proc. Nat. Acad. Sci. USA* **90,** 4596–4600.
11. Mallow, E. B., Harris, A., Salzman, N., Russell, J. P., DeBerardinis, J. R., Ruchelli, E., and Bevins, C. L. (1996) Human enteric defensins: gene structure and developmental expression. *J. Biol. Chem.* **271,** 4038–4045.
12. Tarver, A. P., Clark, D. P., Diamond, G., Cohen, K. M., Erdjument-Bromage, H., Jones, D. E., Sweeney, R., Wines, M., Hwang, S., Tempst, P., and Bevins, C. L. (1997) Enteric expression of a novel epithelial antibiotic peptide, in press.
13. Denhardt, D. T. (1966) A membrane filter technique for the detection of complementary DNA. *Biochem. Biophys. Res. Commun.* **23,** 641–646.
14. Brown, T. (1994) Dot and slot blotting of DNA onto uncharged nylon and nitrocellulose membranes using a manifold in *Current Protocols in Molecular Biology* (Ausubel, F. M., Brent, R., Kingston, R. E., Moore, D. D., et al., eds.), Wiley and Sons, 2.9.95–2.9.20.
15. Sambrook, J., Fritsch, E. F., and Maniatis, T. (1989) *Molecular Cloning: A Laboratory Manual, 2nd ed.* Cold Spring Harbor Laboratory Press, Cold Spring Harbor, NY.
16. Brown, T. (1994) Hybridization analysis of DNA blots, in *Current Protocols in Molecular Biology* (Ausubel, F. M., Brent, R., Kingston, R. E., Moore, D. D., et al., eds.), Wiley and Sons, 2.10.1–2.10.16.
17. Anderson, M. L. M. and Young, B. D. (1985) Quantitative filter hybridisation in *Nucleic Acid Hybridisation, A Practical Approach* (Hames, B. D. and Higgins, S. J., eds.) IRL Press, Oxford, pp. 73–111.
18. Meinkoth, J. and Wahl, G. (1984) Hybridization of nucleic acids immobilized on solid supports. *Analyt. Biochem.* **138,** 267–284.
19. Lathe, R. (1985) Synthetic oligonucleotide probes deduced from amino acid sequence data. Theoretical and practical considerations. *J. Mol. Biol.* **183,** 1–12.
20. Wetmur, J. G. (1991) DNA probes: applications of the principles of nucleic acid hybridization. *Crit. Rev. in Biochem. and Mol. Biol.* **26** 227–259.
21. Britton, R. J. and Davidson, E. H. (1985) Hybridisation strategy, in *Nucleic Acid Hybridisation, A Practical Approach* (Hames, B. D. and Higgins, S. J., eds.), IRL, Oxford, pp 1–15.

22. Wolf, S. F., Haines, L., Fisch, J., Kremsky, J. N., Dougherty, J. P., and Jacobs, K. (1987) Rapid hybridization kinetics of DNA attached to submicron latex particles. *Nucl. Acids Res.* **15,** 2911–2926.
23. Hamaguchi, K. and Geiduschek, E. P. (1962) The effect of electrolytes on the stability of the deoxyribonucleate helix. *J. Am. Chem. Soc.* **84,** 1329–1338.
24. Schildkraut, C. and Lifson, S. (1965) Dependence of the melting temperature of DNA on salt concentration. *Biopolymers* **3,** 195–208.
25. Breslauer, K. J., Frank, R., Blöcker, H., and Marky, L. A. (1986) Predicting DNA duplex stability from the base sequence. *Proc. Natl. Acad. Sci. (USA)* **83,** 3746–3750.
26. McConaughy, B. L., Laird, C. D., and McCarthy, B. J. (1969) Nucleic acid reassociation in formamide. *Biochemistry* **8,** 3289–3295.
27. Casey, J. and Davidson, N. (1977) Rates of formation and thermal stabilities of RNA:DNA and DNA:DNA duplexes at high salt concentrations. *Nucl. Acids Res.* **4,** 1539–1552.
28. Hutton, J. R. (1977) Renaturation kinetics and thermal stability of DNA in aqueous solutions of formamide and urea. *Nucl. Acids Res.* **4,** 3537–3555.
29. Wong, D. M., Weinstock, P. H., and Wetmur, J. G. (1991) Branch captur reactions: displacers derived from assymetric PCR. *Nucl. Acids Res.* **19,** 2251–2259.
30. Hall, T. J., Grula, J. W., Davidson, E. H., and Britten, R. J. (1980) Evolution of sea urchin non-repetitive DNA. *J. Mol. Evol.* **16,** 95–110.
31. Hyman, R. W., Brunovskis, I., and Summers, W. C. (1973) DNA base sequence homology between coliphages T7 and phiII and between T3 and phiII as determined by heteroduplex mapping in the electron microscope. *J. Mol. Biol.* **77,** 189–196.
32. Wallace, R. B., Shaffer, J., Murphy, R. F., Bonner, J., Hirose, T., and Itakura, K. (1979) Hybridization of synthetic oligonucleotides to ΦX 174 DNA: the effect of single base pair mismatch. *Nucl. Acids Res.* **6,** 3543–3557.
33. Ikuta, S., Takagi, K., Wallace, R. B., and Itakura, K. (1987) Dissociation kinetics of 19 base paired oligonucleotide-DNA duplexes containing different single mismatched base pairs. *Nucl. Acids Res.* **15,** 797–811.
34. Brown, T., Leonard, G. A., Booth, E. D., and Kneale, G. (1990) Influence of pH on the conformation and stability of mismatch base-pairs in DNA. *J. Mol. Biol.* **212,** 437–440.
35. Woodson, S. A. and Crothers, D. M. (1988) Structural model for an oligonucleotide containing a bulged guanosine by NMR and energy minimization. *Biochemistry* **27,** 3130–3141.
36. Wetmur, J. G. and Davidson, N. (1968) Kinetics of renaturation of DNA. *J. Mol. Biol.* **31,** 349–370.
37. Bonner, T. I., Brenner, D. J., Neufeld, B. R., and Britten, R. J. (1973) Reduction in the rate of DNA reassociation by sequence divergence. *J. Mol. Biol.* **81,** 123–135.
38. Beltz, G. A., Jacobs, K. A., Eickbush, T. H., Cherbas, P. T., and Kafatos, F. C. (1983) Isolation of multigene families and determination of homologies by filter hybridization methods. *Meth. Enzymol.* **100,** 266–285.

III

Assay Systems for Studying Antibacterial Peptides

12

Designer Assays for Antimicrobial Peptides

Disputing the "One-Size-Fits-All" Theory

Deborah A. Steinberg and Robert I. Lehrer

1. Introduction

Some travelers believe that the difficulties of a voyage become its most pleasurable memories. Others doubt this maxim or believe that it operates only in retrospect, and take a guidebook and compass along. This chapter is written for the second type of traveler.

The problems experienced in attempting to discover antimicrobial peptides can vary, but the following are typical:

1. The extracts or secretions used as starting materials for peptide identification and purification will be available in very limited amounts.
2. The starting materials will contain numerous molecules in addition to those of interest.
3. Some of these bystander molecules, such as mucins or nucleic acids, may bind peptides and mask their antimicrobial activity in crude extracts or may cause anomalous behavior during gel chromatography or ultrafiltration.
4. Crude acid extracts of cells and tissues (typical starting materials) contain other potently antimicrobial components such as histones, histone fragments, ribosomal binding proteins, and so on.
5. Some antimicrobial peptides have a relatively narrow spectrum, making the choice of target organisms important.
6. The activity of some antimicrobial peptides will vary in response to the pH and the concentration of mono- or divalent cations in the assay medium.

Assays intended for use during the discovery stages of antimicrobial peptide research can differ distinctly from assays designed for routine application by

From: *Methods in Molecular Biology, Vol. 78: Antibacterial Peptide Protocols*
Edited by: W. M. Shafer, Humana Press Inc., Totowa, NJ

clinical microbiology laboratories. Discovery-stage assays must be sensitive and exceedingly sparing of material. In contrast, clinical laboratory assays should be cost- and labor-sparing and susceptible to automation. Although discovery-stage assays will often be used in a qualitative mode, especially during early stages of the process, some of them can also provide highly quantitative information once purified peptides are available. As larger amounts of the materials become available via chemical or microbiological syntheses, cost considerations may dictate a switch to assays more like those used in clinical laboratories. This chapter presents a suite of assays of proven utility that can be used for diverse antimicrobial peptides, including defensins, protegrins, tachyplesins, magainins, cecropins, and so on; these peptides are described in other chapters of this volume.

1.1. Radial Diffusion Assay

This method is highly sensitive and consumes minimal amounts of the preparations being tested. It uses microbial target cells that were grown to approximately mid log phase before being entrapped within thin gels that contain dilute trypticase soy broth, 1% (w/v) agarose and 10 m*M* phosphate buffer. Agarose (not agar) is used in this gel to avoid electrostatic interactions between antimicrobial peptides and the polyanionic components of standard agar. Peptides are introduced into small (e.g., 3-mm diameter) sample wells, from which they diffuse radially into the gels. Because the limited nutrient content of this underlay gel will only allow 6–8 bacterial doublings, a nutrient-rich top agar is poured after 3 h of incubation, so that additional colony development can occur. Using target cells entrapped in gels also prevents microbial aggregation, an event that can confound standard colony counting assays.

1.2. Microbroth Dilution Assay

The microbroth dilution method for susceptibility testing offers several advantages over the radial diffusion assay. It accommodates larger numbers of samples and is more amenable to automation. Although the use of serial two-fold dilutions limits the precision of minimal effective concentration determinations, the actual data analysis is direct and simple. Unfortunately, the method requires approx 10 times more peptide than the radial diffusion assay.

The microbroth dilution method is adapted from the radial diffusion assay previously reported. One key step is combining microorganisms and peptide in a defined, minimal nutrient buffer system that minimizes interference with the peptide's biological activity. A second critical step is the addition of 0.1% (w/v) human serum albumin (HSA) to the peptide diluent to minimize adsorption of peptide to the container.

1.3. Modified NCCLS Microbroth Dilution Assay

Addition of our peptides to Mueller Hinton Broth (MHB), the medium most often recommended by the National Committee for Clinical Standards (NCCLS) for MIC determinations with bacteria, resulted in precipitation at peptide concentrations >128 µg/mL *(1)*. Since the NCCLS requires stock solutions be prepared in medium at 512 µg/mL, where the risk of precipitation is great, serial twofold dilutions prepared from such stocks would have less peptide than calculated, resulting in erroneously high MIC estimates. To overcome this problem, we prepare concentrated solutions (10X) of the peptides in 0.01% (v/v) acetic acid containing 0.1% (w/v) HSA and dilute 1:10 into MHB containing the microorganism.

1.4. Gel Overlay Assay

This simple but powerful procedure is especially useful during early stages of antimicrobial peptide purification. It subjects samples (typically crude extracts) to electrophoresis on polyacrylamide minigels that are rinsed and placed on top of agarose gels that contain viable bacteria. After 3 h, the PAGE gels are removed and a nutrient-rich top agar is poured. After overnight incubation to allow surviving bacteria to form microcolonies, the plates are examined for clear (bacteria-free) zones whose number and position provide important information about the responsible effector molecules.

2. Materials

2.1. Radial Diffusion Assay

2.1.1. Microorganisms

Esherichia coli ML-35p contains a β-lactamase encoding plasmid, pBR322, and is maintained on trypticase soy agar supplemented with 100 µg/mL of ampicillin *(2)*. *Listeria monocytogenes* EGD was a gift from Ralph van Furth (University of Leiden, the Netherlands) and has been maintained on trypticase soy agar with 5% sheep blood (Clinical Standards Laboratories, Rancho Dominguez, CA).

In addition to the above bacteria, we have also used this radial diffusion assay for *Staphylococcus aureus, S. epidermidis, Enterococcus faecalis, E. faecium*, Group B streptococcus, *Pseudomonas aeruginosa, Klebsiella pneumoniae, Salmonella typhimurium*, other *E. coli* strains, and *Candida albicans*. When other bacteria are used, the OD/CFU relationship should be determined and used to adjust the formula appropriately. Modifications of the assay designed for use with mycobacteria ***(3)*** and *Neisseria gonorrhoeae* ***(4)*** were recently published.

2.1.2. Media and Reagents

1. Trypticase soy broth (TSB, Difco, Detroit, MI): Full-strength broth contains 30 g of powdered medium per liter of deionized water. It is dispensed into 125-mL bottles, autoclaved for 20 min at 121°C, and stored at room temperature.
2. Phosphate buffer: Stock solutions, each 100 m*M*, of monobasic sodium phosphate ($NaH_2PO_4 \cdot 2H_2O$, Fisher) and dibasic sodium phosphate ($Na_2HPO_4 \cdot 7H_2O$ Fisher) are prepared by dissolving 15.6 g of the monobasic or 26.8 g of the dibasic salt in 1 L of deionized distilled water. To prepare 100 m*M* phosphate buffer at pH 7.4 or 6.5 (or as desired), the monobasic and dibasic phosphate solutions are mixed together while stirring and monitoring the pH. The resulting buffers can be either autoclaved at 121°C for 20 min or filter sterilized (0.45-µm filter), and then stored at room temperature. Although their preparation is not described here, one may also perform radial diffusion assays with various other buffers, including TRIS, HEPES, MOPS, citrate-phosphate, and acetate.
3. Agarose (Sigma): The use of a low electroendosmosis (EEO)-type agarose in place of standard agar is *critical* to limit electrostatic interactions between positively charged antimicrobial peptides and sulfated moieties of the agaropectin component of standard agar.
4. Underlay gels: Mix 50 mL of 100 m*M* sodium phosphate buffer with 5 mL full-strength trypticase soy broth in a 1000-mL Pyrex beaker, add 5 g agarose (Sigma), and bring the volume to 500 mL with deionized distilled water. Adjust the final pH to 7.4 or 6.5 with 1*N* NaOH or HCl, place the suspension on a hot plate, and stir it under heat until the agarose dissolves. Using a 50-mL serological pipet or a graduated cylinder, dispense 50-mL aliquots into the 125-mL bottles, autoclave them at 121°C for 20 min, and store the sterilized media at room temperature. Before use, the solidified medium should be fluidized (we use a microwave oven) and placed into a water bath maintained at 42°C. When constituted as above, the underlay gels contain 10 m*M* sodium phosphate buffer, 0.3 mg/mL of TSB powder, and 1% (w/v) of agarose. For certain uses, we have modified the underlay gel to include 100 m*M* NaCl, 2.5–5% normal or heat-inactivated human serum, tissue culture media such as RPMI-1640, surfactants such as Tween-20, or physiological concentrations of divalent ions (Mg^{+2}, Ca^{+2}).
5. Overlay agar: This agar contains 60 g (twice the customary amount) of trypticase soy broth (TSB, Difco) and 10 g of agarose (Sigma) per liter of deionized water. The suspension is placed in a Pyrex beaker and stirred on a hot plate until the agarose dissolves. It is dispensed in 50-mL aliquots into 125-mL bottles that are autoclaved for 20 min at 121°C. The sterilized overlay agar is stored at room temperature. Prior to use, it is heated in a microwave oven to redissolve the agarose, placed in a 42°C water bath to keep it molten, and transferred in 10-mL aliquots to 15-mL centrifuge tubes shortly before they are used to pour the overlay gels. It is also acceptable to use conventional trypticase soy or Sabouraud's

agar (for *C. albicans*) for the overlay gels, if these are reconstituted at twice the concentration indicated on the label.

6. 10 m*M* sodium phosphate buffer, pH 7.4: This is prepared by diluting 100 m*M* sodium phosphate buffer, sterilized and maintained at room temperature. It is chilled in an ice bucket before being used to wash bacteria.

2.1.3. Hardware

1. Square plastic Petri dishes, 10 × 10 × 1.5 cm (Nunc, Baxter Scientific Products).
2. Spectrophotometer (e.g., Beckman, Fullerton, CA).
3. Single-channel adjustable pipetman, Models P-20 and P-200 (Rainin, Woburn, MA).
4. Sterile polypropylene tips (E & K Scientific Products, Saratoga, CA).
5. Gel-leveling table (Research Products, Mount Prospect, IL).
6. Gel punch, stainless steel, custom made (4.5 × 0.3 cm).
7. Conical capped clear microcentrifuge tubes, 1.5-mL capacity (RingLock, United Scientific Products, San Leandro, CA).
8. Water bath, stationary (e.g., Fisher).
9. Water bath, shaking (e.g., New Brunswick Model G76D, Fisher).
10. Incubator, 37°C.
11. 7X Magnifier (Bausch & Lomb, Fisher).
12. Sterile polystyrene loops ("steriloops," Fisher).
13. Centrifuge tubes, 15-mL (Fisher) and 50-mL (Fisher).
14. Mixer, vortex.
15. Template: A simple template is made by marking a piece of cm-ruled graph paper with a regularly spaced 4 × 4 (16 samples) or 5 × 5 (25 samples) array of 3-mm circles.
16. Bleach trap: Approximately 400 mL of industrial bleach (sodium hypochlorite) is added to a 2-L side-arm flask (Fisher) fitted with a one-hole rubber stopper and tubing. The unit is attached to a vacuum source via the stopper, and to a glass Pasteur pipet via the side arm.
17. 125-mL bottles (Fisher).
18. SpeedVac centrifuge (Savant Instruments, Holbrook, NY).
19. Hot plate/magnetic stirrer (Fisher).
20. Serological pipets, 50-mL (Fisher).
21. Some commercially available antimicrobial peptides (for use as standards): Cecropin P1 (Sigma); Magainin-1 (Bachem, Torrance, CA); Tachyplesin-1 (Bachem); polymyxin B sulfate (Sigma); Protamine sulfate (Sigma).

2.2. Microbroth Dilution Assay

2.2.1. Microorganisms

E. coli ML-35p is described **Subheading 2.1.1.** *P. aeruginosa* ATCC 9027 and methicillin-resistant *S. aureus* (MRSA) ATCC 33591 were obtained from the American Type Culture Collection, Rockville, MD.

2.2.2. Media and Reagents

1. Trypticase soy agar (TSA; Becton-Dickinson, Cockeysville, MD); dissolve 40 g in 1 L of deionized water, autoclave 121°C for 20 min.
2. Trypticase soy broth (TSB; Becton-Dickinson): dissolve 30 g in 1 L deionized water, autoclave 121°C for 20 min, and store at room temperature.
3. 2X Trypticase soy broth (2X TSB): Dissolve 60 g in 1 L deionized water, autoclave 121°C for 20 min, and store at room temperature.
4. Glycerol (20% v/v): Mix 20 mL glycerol with 80 mL deionized water, filter sterilize with a 0.45-μ filter, and store at room temperature.
5. 100 m*M* monobasic phosphate buffer: Dissolve 13.7 g sodium phosphate monobasic (Fisher) in 1 L deionized water, filter sterilize with a 0.45-μ filter, and store at room temperature.
6. 100 m*M* dibasic phosphate buffer: Dissolve 14.2 g sodium phosphate dibasic (Fisher) in 1 L deionized water, filter sterilize with a 0.45-μ filter, and store at room temperature.
7. 10 m*M* Phosphate buffer, pH 7.4, 100 m*M* NaCl: Combine 15 mL dibasic phosphate buffer (100 m*M*), 5 mL monobasic phosphate buffer, 4 mL 5*M* NaCl, and 176 mL deionized water, adjust pH if necessary, filter sterilize with a 0.45-μ filter, and store at room temperature.
8. 100 m*M* Phosphate buffer, pH 6.5: Combine 40 mL dibasic phosphate buffer (100 m*M*), 160 mL monobasic phosphate buffer. Adjust pH if necessary, filter sterilize with a 0.45-μ filter, and store at room temperature.
9. Liquid testing medium (LTM): Aseptically combine the following sterile ingredients: 10 mL 100 m*M* phosphate buffer, pH 6.5, 1.0 mL TSB, 2 mL 5*M* NaCl, and 87 mL deionized water. Store at room temperature.
10. 0.01% Acetic acid: Mix 10 μL of acetic acid into 100 mL of sterile deionized water.
11. Peptide diluent: Add 0.1 g 0.1% human serum albumin (HSA; Bayer-Miles, Kankakee, IL) to 100 mL 0.01% acetic acid and filter sterilize. Store at room temperature.

2.2.3. Hardware

1. Sterile inoculation loops.
2. Erlenmeyer flasks (50 and 250 mL); sterilize by autoclaving.
3. Spectrophotometer (LKB Ultrospec II).
4. Sterile 96-well polypropylene microtiter plates (Costar, Cambridge, MA).
5. Sterile polypropylene tips (Rainin).
6. Micropipettor: 10–200 μL range; P20 and P200.
7. Multichannel micropipettors: 10–50 μL and 50–200 μL range.
8. Sterile pipets: 1 mL, 10 mL.
9. Sterile 50-mL polypropylene centrifuge tubes.

10. Vortex mixer.
11. Sterile V-well troughs (Costar, Cambridge, MA).
12. Sterile 24-well microtiter plates (Costar).
13. Microtiter Plate warmer, 37°C (Thermolyne, Dubuque, IA).

2.3. Modified NCCLS Microbroth Dilution Assay

2.3.1. Microorganisms (see **Subheading 2.2.1.**)

2.3.2. Media and Reagents

1. Mueller Hinton Broth (MHB; Becton-Dickinson). Dissolve 22 g in 1 L deionized water; autoclave at 121°C for 20 min.

2.4. Gel Overlay Assay

2.4.1. Microorganisms (see **Subheading 2.1.1.**)

2.4.2. Media and Reagents

1. Chemicals: Acetic acid (Fisher), acrylamide (Gibco-BRL, Gaithersburg, MD), ammonium persulfate (Fisher), *N,N′*-methylene-bisacrylamide (Gibco-BRL), Coomassie brilliant blue R-250 (Sigma), formaldehyde (Fisher), *N,N,N′,N′* tetramethylethylenediamine (TEMED, Fisher), methyl green (MCB), urea (Fisher).
2. Staining solution: Coomassie brilliant blue R-250 (0.1 g) dissolved in methanol (27 mL) plus water (63 mL) plus 37% formaldehyde (15 mL).
3. Destaining solution. Methanol (100 mL) plus water (150 mL) plus 37% formaldehyde (4 ML).
4. Other: *see* **Subheading 2.1.2.**

2.4.3. Hardware

1. Minigel unit: "Mighty Small II" SE 250 unit (Hoefer Instruments, San Francisco, CA).
2. Power supply (Fisher).

3. Methods

3.1. Radial Diffusion Assay

To prepare organisms for the assay:

1. A single colony is picked with a polystyrene loop, transferred to a 125-mL capacity bottle that contains 50 mL of TSB, and incubated at 37°C in a shaking water bath for 18–24 h.
2. An aliquot (50 µL of *E. coli* or 500 µL of *L. monocytogenes*) of the resulting stationary phase cultures is transferred to 50 mL of fresh TSB and incubated for 2.5 h at 37°C in a shaking water bath.

3. This subculture is transferred to a 50-mL conical tube and centrifuged for 10 min at 4°C at approx 880*g*.
4. The bacterial pellet is washed once (10 min, 4°C, approx 880*g*) with 10 mL of cold sterile 10 m*M* sodium phosphate buffer, pH 7.4, and resuspended in 5 mL of the same cold buffer.
5. One milliliter is removed to measure its optical density at 620 nm. From this information, the concentration of bacteria in the remaining 4 mL is calculated from the following formula, which is applicable to either organism: CFU/mL = $0D_{620} \times 2.5 \times 10^8$. From this calculation, the volume of washed bacterial suspension that contains 4×10^6 CFU (the inoculum used for each underlay) can be determined.
6. A 10-mL aliquot of the sterile, molten underlay agar (maintained at 42°C as described above) is transferred to a 15-mL conical plastic centrifuge tube, inoculated with 4×10^6 CFU of washed bacteria, vortexed vigorously for 15 s, and then poured into a $10 \times 10 \times 1.5$ cm square dish on a leveling table (very important!), where it gels in less than 2 min. When multiple plates are required for an experiment, they can be poured serially in this manner, since the leveling table accommodates three plates at a time. We typically offset the position of the lowest right well or punch an extra well in the lower right corner to indicate the correct orientation of the plates. In addition, we label both the top and bottom halves of each Petri dish.
7. After the underlay gel has set, the plates are placed over a graph paper template and a 4×4 or 5×5 array of wells is punched. The central plugs are removed by suction, using a Pasteur pipet attached to a bleach trap.
8. Five-microliter aliquots of the various samples to be tested, typically diluted in 0.01% acetic acid, are added to each well in turn. The plates are covered, turned gel-side up, and incubated for 3 h in a 37°C incubator.
9. Next, each underlay gel is covered with a 10-mL overlay of nutrient-rich overlay agar. As soon as the overlay gel solidifies, the plates are recovered and placed gel-side up and incubated at 37°C overnight.
10. The following morning, the plates are removed, and 10 mL of a disinfecting solution (e.g., 5% acetic acid in 25% methanol) is applied to the agar surfaces for at least 20 min before the zone diameters are measured with a magnifier of ×7 and recorded.
11. Alternatively, we have stained the gels for 24 h with a dilute solution of Coomassie brilliant blue R-250 (dye, 2 mg; methanol, 27 mL; water, 63 mL; and 37% formaldehyde, 15 mL) and measured the zone diameters after decanting the spent solution. Gels can also be dried and retained indefinitely as experimental records. In this case, the gels should be conditioned by adding 10 mL of 10% acetic acid with 2% dimethylsulfoxide. After 10 min, decant this conditioning solution. Such gels dry without cracking in approx 60 min on a Bio-Rad Model 224 slab gel drier (Bio-Rad, Berkeley, CA).

Table 1
Data Analysis and Calculations

Concentration, μg/mL	Tachyplesin-1	Polymyxin *B*	Protamine sulfate
250	94[a]	140	50
79.1	68	118	30
25	42	92	10
7.91	18	60	0
2.5	6	38	0
0.79	0	16	0
0.25	0	0	
Correlation coefficient	0.992	0.998	1.00
X-intercept	2.45 μg/mL	0.43 μg/mL	14.1 μg/mL

[a]Zone diameter given in units (10 U = 1 mm).

3.1.1. Calculating the Results

Experimental data from an assay done with *E. coli* ML-35p and three commercially available antimicrobial peptides are shown in **Table 1**. In this experiment, the underlay gel was supplemented with 100 m*M* NaCl. The diameters of the clear zones were measured to the nearest 0.1 mm. Then, the diameter of the well (3.0 mm) was subtracted and the difference was multiplied by 10 to convert the zone diameter to units (10 U = 1 mm). The correlation coefficients and X-intercepts were obtained from linear regression analyses (wherein X = $\log_{10}$ peptide concentration and Y= zone diameter, in units) that was performed with a simple scientific calculator. When performing such analyses, only nonzero zone values should be included. Whereas both tachyplesin and protamine sulfate gave completely clear zones, the polymyxin B zones had only a partial (approx 90%) reduction in colonial density at all concentrations tested, making it a less satisfactory standard.

3.1.2. Graphing the Results

After the zone diameters have been converted to units, these values are graphed on semilogarithmic coordinates against the peptide concentrations that had been introduced into the wells. We most often use a "half-log" dilution series that starts at 250 μg peptide/mL and goes down to 0.25 μg/mL. To prepare this, six serial 3.16-fold dilutions of the 250 μg/mL concentration are prepared (3.16 is the square root of 10). We used the SigmaPlot® Graphics program (Jandel San Rafael, CA) to draw the graph and its regression lines.

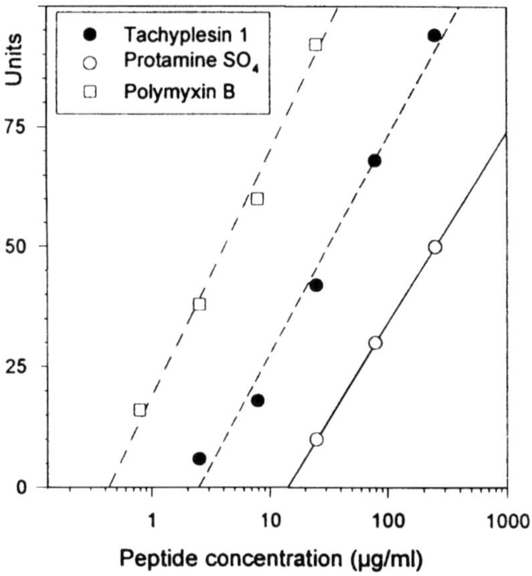

Fig. 1. Radial diffusion assay. The data shown in **Table 1** have been graphed. Note the logarithmic scale of the *x*-axis.

The minimal effective concentration (equivalent to the X-intercept) can be read directly from the graph and should be the same as the values that were obtained by linear regression analysis as described in **Subheading 3.1.1.** (*see* **Fig. 1**).

3.2. Microbroth Dilution Assay

3.2.1. Microorganisms

1. Bacteria are cultured on TSA.
2. Isolated colonies are transferred into TSB (10 mL in a sterile 50-mL Erlenmeyer flask) using a sterile, disposable loop, and the flask is incubated at 37°C in a shaking water bath (200 rpm) for 16–18 h.
3. Broth cultures are diluted 1:1 with 20% sterile glycerol and stored as 1.0 mL aliquots at –80°C. For daily inocula, liquid should be transferred from a thawed vial using a sterile loop and then spread onto the surface of TSA slants.
4. The screw-capped tubes are then incubated overnight and stored at 4°C for up to 1 mo.

3.2.2. Preparation of Inoculum

1. Remove the cap from tube and lightly touch a sterile loop to the area of heavy growth on the TSA slant.
2. Inoculate 10 mL of TSB in a 50-mL flask, swish vigorously to release bacteria from the loop, and place the flask in a shaking water bath (200 rpm) for 18 h at 37°C.

3. Dilute the overnight culture 1:20 (50 µL culture plus 950 µL TSB). Using TSB as a reference blank, measure the absorbance of your diluted culture at 600_{nm}. The $A_{600\ nm}$ (1:20) should be between 0.1 and 0.4.
4. Dilute the overnight culture 1:1000 by adding 50 µL of the overnight culture to 50 mL of fresh TSB in a 250-mL Erlenmeyer flask.
5. Incubate the diluted culture in a shaking water bath at 37°C, 200 rpm, for approx 2–3 h until the absorbance of the undiluted culture is between 0.200 and 0.400.
6. Centrifuge 25 mL of the log-phase culture at 3000*g* at 4°C for 10 min. Decant the supernatant, add 25 mL of sterile phosphate buffer (10 m*M*, pH 7.4, 100 m*M* NaCl), and resuspend the pellet by vortexing.
7. Centrifuge the suspension at 3000*g* at 4°C for 10 min. Decant the supernatant and resuspend the pellet with 5 mL sterile phosphate buffer (10 m*M*, pH 7.4, 100 m*M* NaCl).
8. Measure the absorbance of the undiluted culture. If the absorbance is >0.5, dilute the organism with sterile phosphate buffer (10 m*M*, pH 7.4, 100 m*M* NaCl) until the absorbance is between 0.100 and 0.500.
9. Determine the number CFUs/mL of suspension by preparing 10-fold serial dilutions in saline (0.87%) and spreading 100 µL of a 10^{-4}, 10^{-5}, and a 10^{-6} onto three separate TSA plates. Incubate overnight and count the number of colonies. An accurate determination requires approx 30–300 colonies on a plate. Once you have determined the number of CFUs for each organism, you can calculate the amount of suspension to add to LTM as follows:

$$\text{(Measured } A_{600\ nm}) \times \frac{\text{(no. CFUs/mL)}}{A_{600\ nm} \text{ of } 0.2} = \text{no. CFUs/mL of suspension}$$

   For the strains reported here, we have determined the CFUs/mL as shown in **Table 2**.
10. Approximately 10 mL of diluted cell suspension (4×10^5 CFUs/mL) is required for each microtiter plate (100 µL per well × 96 wells = 9.6 mL). Calculate the total volume required for entire assay (e.g., 5 plates × 10 mL per plate = 50 mL). Prepare the solution of LTM as follows and keep on ice until ready for use:

$$\frac{\text{(no. CFUs/mL of suspension)} \times (X \text{ mL})}{\text{(50 mL of LTM)}} = 4 \times 10^5 \text{ CFUs/mL}$$

3.2.3. Preparation of Stock Solutions of Peptides

1. Weigh out approx 1 mg of each peptide to be tested into a sterile polypropylene cryovial (1.8 mL).
2. Add sufficient 0.01% acetic acid to make a stock solution of 1280 µg/mL.
3. Aliquot 100 µL of stock solution into multiple vials and store tightly sealed at –80°C.

Table 2
Relationship of CFU to Absorbance

Microorganism	(no. CFUs/mL) 0.2 $A_{600\ nm}$
E. coli	8×10^7
P. aeruginosa	7.8×10^7
MRSA	2×10^7

4. From a single vial, prepare serial twofold dilutions in peptide diluent (0.1% HSA in 0.01% acetic acid). Using the multichannel pipettor, dispense 50 µL of peptide diluent (0.1% HSA in 0.01% acetic acid) into wells 2–12. Transfer stock peptide solution (1280 µg/L) to well Al. Mix well by triturating three times, then transfer 50 µL of stock peptide solution to well A2 containing 50 µL of diluent = 640 µg/mL. After triturating three times with the pipettor, transfer 50 µL from well A2 to A3 = 320 µg/mL. Repeat process until all 12 wells have been prepared. The last well will contain 100 µL of a solution = 0.6 µg/mL. If several peptides are to be tested, you may use a multichannel pipettor to transfer several wells at one time.

3.2.4. Preparation of Assay Plates

1. Dispense 100 µL of the resuspended log-phase cell suspension (4×10^5 CFUs/mL) into each well of a 96-well polypropylene microtiter plate.
2. Add 11 µL of each peptide dilution (A1–A12) to each well containing the cell suspension. Mix well by triturating three times.
3. Repeat process two times to provide three sets of serial dilutions for each peptide tested.
4. Incubate the plates at 37°C for 3 h.
5. Add 100 µL of 2X TSB to each well, mix, and reincubate at 37°C for 16–18 h without aeration.
6. Examine the plates and score each well for turbidity. Often, the MRSA bacteria will settle out and form a pellet of growth at the bottom of the well. This strain can be scored by placing the microtiter plate on a stand and examining the bottom using a tilted mirror.
7. The last well in the series without any visible growth is used to calculate the MCB (minimum concentration for inhibition of growth in broth) medium. If the last clear well differs in the three series, the MCB is calculated by averaging the concentration of peptide in each of the three wells.
8. To determine the bactericidal activity of the peptides, transfer 10 µL from the MCB wells, 2X the MCB, and 4X the MCB to TSA (1.5 mL in each well of a 24-well plate: minimizes the quantity of plates and crosscontamination).
9. The concentration that does not result in any detectable growth on TSA is considered the minimum bactericidal concentration (MBC). If the well differs in the

three series, the MBC is calculated by averaging the concentration of peptide in each of the three wells.

3.2.5. Kinetic Bactericidal Assay

To compare the rate at which peptides kill bacteria, one can perform a relatively simple kinetic assay. Follow **Subheading 3.2.1.–3.2.3.** in the microbroth dilution assay for preparation of microorganism, inocula, and peptides. Then:

1. Dispense 200 µL of the resuspended log-phase cell suspension (4×10^5 CFUs/mL) into each well of a 96-well polypropylene microtiter plate solution.
2. At time (T) = 0, add 22 µL of a peptide solution to the first well (A1) and mix by triturating three times.
3. Stagger the addition of the next peptide solution by 30 s, add 22 µL of a second concentration to the next well (A2), and mix by triturating three times.
4. Repeat process until all concentrations have been added. We use a fourfold series of dilutions (i.e., 1280, 320, 80, 20, 5 µg/mL diluted 1:10 into each well) for comparative kill curves.
5. For a control, add 22 µL of 0.01% acetic acid to one well.
6. At T = 15 min, mix well A1 by triturating three times and transfer 20 µL to a sterile Petri dish (100 × 15 mm).
7. Quickly add 20 mL of tempered (50°C) TSA and gently swirl plate to mix.
8. After the agar has solidified, invert the plate and incubate at 37°C for 18 h.
9. For the control well containing acetic acid, you must dilute the sample 1:100 by transferring 20 µL into 2.0 mL of LTM, then transfer 50 µL of the dilution into a Petri dish to obtain accurate determinations of CFUs.
10. Repeat the entire process for all peptide concentrations and the control well at T = 30, 60, 120, and 240 min.
11. Count the number of CFUs per plate and estimate the reduction in CFUs by each peptide.

The peptide must reduce the CFUs by ≥1 log (i.e., 800 CFUs per plate) to assess a bactericidal effect in this assay method. Although such numbers are higher than recommended for accurate CFU determinations (30–300 CFUs/plate), log changes in recoverable CFUs indicate significant bactericidal efficacy. The data are plotted as the log of the fractional survival (CFUs/mL treated sample/CFUs control at each time point) vs peptide concentration.

3.3. Modified NCCLS Microbroth Dilution Assay

3.3.1. Microorganisms (see **Subheading 3.2.1.**)

3.3.2. Preparation of Inoculum

1. Remove the cap from tube and lightly touch a sterile loop to the area of heavy growth on the TSA slant.

2. Inoculate 10 mL of MHB in a 50-mL flask, swish vigorously to release bacteria from the loop, and place the flask in a shaking water bath for 18 h at 37°C, 200 rpm.
3. Dilute the overnight culture 1:20 (50 µL culture plus 950 µL MHB). Using MHB as a reference blank, measure the absorbance of your diluted culture at 600_{nm}. The $A_{600\ nm}$ (1:20) should be 0.1–0.4.
4. Dilute the overnight culture 1:10,000 (to approx 4×10^5 CFUs/mL) by adding 5 µL of the overnight culture to 50 mL of fresh MHB in a 250 mL Erlenmeyer flask. You should deterine the ratio of CFUs/$A_{600\ nm}$ for each organism and adjust the dilution factor as necessary to achieve the recommended CFUs/mL.

3.3.3. Preparation of Stock Solutions of Peptides (see **Subheading 3.2.3.**)

3.3.4. Preparation of Assay Plates

1. Dispense 100 µL of the freshly diluted overnight culture into each well of a 96-well polypropylene microtiter plate.
2. Add 11 µL of each peptide dilution (A1–A12) to each well containing the cell suspension. Mix well by triturating three times.
3. Repeat process two times to provide three sets of serial dilutions for each peptide tested.
4. Incubate the plates at 37°C for 16–18 h without aeration.
5. Examine the plates and score each well for turbidity. Often, the MRSA bacteria will settle out and form a pellet of growth at the bottom of the well. This strain can be scored by placing the microtiter plate on a stand and examining the bottom using a tilted mirror.
6. The last well in the series without any visible growth is used to calculate the MIC (minimum inhibitory concentration) for inhibition of growth. If the last clear well differs in the three series, the MIC is calculated by averaging the concentration of peptide in each of the three wells.
7. To determine the bactericidal concentration of the peptides, transfer 10 µL from the MIC wells, 2X the MIC, and 4X the MIC to TSA (1.5 mL in each well of a 24-well plate: minimizes the quantity of plates and crosscontamination).
8. The concentration that does not result in any detetctable growth on TSA is considered the minimum bactericidal concentration (MBC). If the well differs in the three series, the MBC is calculated by averaging the concentration of peptide in each of the three wells.

3.4. Gel Overlay Assay

3.4.1. Gel Preparation

The following recipe will suffice to prepare four acid-urea polyacrylamide minigels, each 100 mm wide × 75 mm long × 0.75 mm thick. The gels are polymerized with a 10-lane comb in place, and the ammonium persulfate and

TEMED are removed by prerunning the gels with 5% acetic acid for approx 45–60 min at 150 V prior to use.

Component	Quantity
urea	2.8 g
distilled water	18.0 mL
60% acrylamide + 1.6% *bis*-acrylamide	8.9 mL
acetic acid	43.2%
N,N,N′,N′ tetramethylethylenediamine, 4% (v/v)	5.33 mL
ammonium persulfate, 10%	0.8 mL

3.4.2. Electrophoresis

1. No stacking gel is used. The samples (5 µL) prepared in 3*M* urea with 5% acetic acid, and electrophoresed with 5% acetic acid at 150 V (approx 15 mA/plate), for approx 45 min or until the methyl green tracking dye comes close to the end of the gel.
2. The gels are removed and cut in half from top to bottom. One hemigel is stained with Coomassie blue. The other hemigel, which was loaded with identical samples, is tested for antimicrobial activity after rinsing it once or twice with buffer, 10 min each time, to remove much of the acetic acid and urea. This can be done in an improvised apparatus made from an empty Fisher micropipet Redi-tip box that contains the perforated plastic platform that formerly held the pipet tips and a teflon-coated, 1 × 4-cm magnetic stirring bar. The perforated platform should be trimmed with a single-edged razor so that it stands about 2 cm tall.
3. The gel is placed on top of the platform and the stir bar beneath it, and the entire assembly is placed on a magnetic stirrer. The first rinse is performed with approx 200 mL of sterile 10 m*M* sodium phosphate buffer containing (optional) 200 µL of 1*M* NaOH. The second rinse, when needed, is performed with 200 mL of sterile, 10 m*M* sodium phosphate buffer. *Caution:* Since peptides can diffuse out of the gel during rinsing, do not prolong this process unduly, especially for small peptides.
4. Once rinsed, the gel is placed on top of an agarose underlay gel containing *E. coli* or *L.* monocytogenes (or other organisms, if desired) and prepared exactly as described for radial diffusion assays, except for lacking sample wells. The exact placement of the PAGE gel's origin can be marked on the underlay gel by making several holes through both with a gel punch. After 3 h of incubation at 37°C (to allow transfer of the electrophoresed polypeptides to the underlay gel), the polyacrylamide gel is removed and replaced by an overlay of double strength nutrient agar that is also prepared exactly as if for radial diffusion assays. The removed PAGE- hemigel can be stained with Coomassie brilliant blue and compared with its stained but untransferred replica hemigel to verify the efficiency of peptide and protein transfer. After incubation for 18–24 h at 37°C, the plates are examined for zones of clearing, which can be enhanced by staining the agarose

gels with dilute Coomassie blue, as described in the radial diffusion assay. The location of the clear zones seen in the underlay should be compared with the banding patterns seen in Coomassie-stained but untransferred gels. If 1 μg of a highly antimicrobial peptide (either protegrin PG-1, tachyplesin TP-1, or defensin NP-1) was placed in the outside lanes of the gel to serve as a positive control and marker, the ratio of the distance migrated from the origin of the unknown peptides to the distance migrated by the standard (we call this number the relative migration) provides a useful and remarkably reproducible attribute of the unknown peptide that can be followed during its subsequent purification.

4. Notes

1. On occasion (*see* **Subheading 3.1.**), a ring-like precipitate may be present at the well's boundary, suggesting that some protein or peptide may have aggregated or precipitated instead of diffusing into the underlay agars. This may be more visible after the plates have been fixed and stained with Coomassie blue as described in **Subheading 3.**
2. Because small antimicrobial molecules can diffuse more rapidly into the underlay agars than larger molecules, they give larger clear zones when tested at equivalent mass or molar concentrations. However, using the X-intercept rather than the zone diameter to interpret activity will make the assay largely independent of the diffusion rates.
3. Especially when the wells are surrounded by large clear zones, the surviving colonies at the edge of the clearing will be larger than those in the control areas around the wells. Most likely, this is owing to the additional nutrient available to these survivors by virtue of the absence of viable organisms between them and the well. Sometimes, depending on the peptide, the assay conditions and the test organism, a completely clear zone around a sample well may be surrounded by a concentric zone of partial clearing owing to a 30–90% reduction in microcolony density. Should this occur, the diameters of both the complete and partial killing zones should be measured and recorded but only the inner-zone diameter (complete clearing) should be used to calculate the X-intercept.
4. It can be instructive to look at radial diffusions plates with a regular or inverted microscope, using a very low power objective lens. We use a Nikon 3.2X objective for this purpose. By placing an grid-ruled micrometer disk (e.g., Fisher) inside one of the eyepiece lenses and calibrating the grid with a stage micrometer (Fisher), the plate can be treated like a hemocytometer and the colonies per mm^3 of underlay agar can be counted using some very simple geometrical formulas for area and volume. We count the colonies in "control" portions of the plate (i.e., those between the sample wells) and also those in areas of where colony count reduction is incomplete.
5. Sometimes, especially when one is just learning the radial diffusion technique, after the overnight incubation, the overlay surface is covered with bacterial colonies that are larger than the microcolonies entrapped in the agarose underlay gel. Generally, these organisms are not contaminants, but are identical to the bacteria

entrapped within the underlay gel. Curiously this "problem" tends to go away once a person has performed the assay for a while. If it should occur, the surface organisms can be washed off with 10 mL of a disinfecting solution, such as 5% acetic acid in 25% methanol.

6. Finally, the choice of test organisms and test conditions will often determine which peptides will emerge from a "grind and find" search for new natural products. One of us (R.I.L.) believes that his repeated encounters with small cystine-rich antimicrobial peptides is a direct consequence of including the fungus, *C. albicans*, among the primary microbial targets in his research.
7. We realize that many of our readers will have "perfectly good agar," filter paper disks, and NCCLS manuals nearby, and, because habit and tradition are comforting, would automatically use these materials for testing antimicrobial peptides. To them we convey our best wishes with the reminder that these excellent tools were never designed for testing antimicrobial peptides, whereas the methods described in this chapter were developed specifically for this purpose.
8. Although the storage conditions (*see* **Subheading 3.2.1.**) described here are acceptable for these strains, some microorganisms do not survive well on slants, e.g., *Enterococcus faecium*, and should be evaluated for viability. Deviations from the expected absorbances of overnight cultures or significant changes in the time it takes the culture to reach log-phase suggests possible contamination or loss of viability.
9. Variation in MCB values (*see* **Subheading 3.2.4.**, **step 7**) can result from erroneous inocula levels. Cells left in the buffered saline for more than an hour may lose viability, even on ice. The presence of NaCl (100 m*M*) in the buffer is critical for *P. aeruginosa*, which appears to lyse in buffer alone. Most important is vigorous mixing of the bacterial suspensions, especially after centrifugation, to disperse any clumps and provide an even inoculum suspension. Each cell suspension should be serially diluted and plated onto TSA to determine the actual inocula used in the experiment.
10. We have shown that the addition of carrier protein (e.g., HSA) can significantly enhance the bioavailability of the peptide, especially at concentrations ≤10 µg/mL. Presumably, HSA binds to surfaces, thereby blocking the nonspecific adsorption of the peptide. However, since different lots of HSA or HSA from different manufacturers exhibit wide variation in their protective effects, it is critical to test each new lot before use in the assay. Recently, we found bovine serum albumin (BSA, Sigma) satisfactory as a less expensive substitute for HSA.
11. For maximum sensitivity, the LTM medium is prewarmed to 37°C prior to addition of the bacterial suspension in the microbroth dilution assay. If the plate is used for bactericidal kinetics, it is kept at 37°C using the plate warmer during the sampling period.
12. Since the outer rows of the microtiter plates (*see* **Subheading 3.3.4.**) tend to dry out during overnight incubation, especially in the Modified NCCLS MIC assay where only 100 µL is present, we use rows B–G for the peptide series and leave A and H for blank controls. Evaporation loss can be minimized using a humidi-

fied incubator or by adding a pan of water to the chamber. When transferring peptide to the liquid in the microtiter plate, it is important to change tips each time to avoid contamination of the stock solutions.

13. Often, MRSA bacteria will form microcolonies or diffuse threads of growth in the bottom of the well, especially as the concentration approaches the MIC. If you find MBCs substantially higher (i.e., ≥10X MIC), it is likely that you missed the faint growth and recorded too low a value for the MIC wells.
14. The most problematic part of this assay is the postelectrophoresis washing step (**Subheading 3.4.2.**). Insufficient washing leaves too much acetic acid in the acrylamide gel, making the entire gel toxic to bacteria in the underlay. If the washing step is too prolonged, the molecules of interest can elute and be lost, especially when of low molecular weight (M_r). We have used the procedures described above successfully with a variety of peptides, α-helical, β-sheet, or proline-rich, ranging in mass from <2 to 8 kDa. Although lysozyme (14 kDa), leukocyte cathepsin G (25–28 kDa), and certain antimicrobial histones (apparent M_r of 31–37 kDa) can also be detected, the ability of the gel overlay assay to detect larger antimicrobial proteins is uncertain. For some antimicrobial peptides, especially those rich in histidine residues, the overlay assay works better when the underlay agars are more acidic (pH 6.0 or 6.5). The acid urea PAGE system is most useful for highly cationic peptides, but any peptide that has a net positive charge in 5% acetic acid (approx pH 2.5), the running buffer, should theoretically enter the gel. The ability of the system to detect polyanionic antimicrobial peptides that are especially rich in aspartic or glutamic acid residues is uncertain.

References

1. *Methods for Dilution Antimicrobial Susceptibility Test for Bacteria that Grow Aerobically, 3rd ed.* Approved Standard M7-A3. The National Committee for Clinical Standards, Villanova, PA.
2. Lehrer, R. I., Barton, A., and Ganz, T. (1988) Concurrent assessment of inner and outer membrane permeabilization and bacteriolysis by multiple-wavelength spectrophotometry. *J. Immunol. Meth.* **108,** 153.
3. Qu, X. D., Harwig, S. S. L., Oren, A., Shafer, W. M., and Lehrer, R. I. (1996) Susceptibility of *Neisseria gonorrhoeae* to protegrins. *Infect. Immun.* **64,** 1240–1245.
4. Miyakawa, Y., Ratnakar, P., Rao, A. G., Costello, M. L., Mathieu-Costello, O., Lehrer, R. I., and Catanzaro, A. (1996) In vitro activity of the antimicrobial peptides human and rabbit defensins and porcine leukocyte protegrin against *Mycobacterium tuberculosis*. *Infect. Immun.* **64,** 926–932.

13

Interaction of Cationic Peptides with Bacterial Membranes

Shafique Fidai, Susan W. Farmer, and Robert E. W. Hancock

1. Introduction

A common feature of cationic peptides is that their site of action is at the membrane due to channel formation, and that they tend to possess strong selectivity towards their target membrane. For example, although moth cecropin and bee melittin are members of the same family of peptides that adopt amphipathic α-helical structures, the cecropins are strongly antibacterial and demonstrate minimal eukaryotic selectivity (i.e., toxicity), whereas melittin is a weak antibacterial compound but a potent toxin. Whereas the basis for selectivity is not completely understood, it has been shown to be due to the size of the transmembrane electrical potential gradient (up to –140 mV in bacterial cytoplasmic membranes compared with about –20 mV or less in eukaryotic membranes) and the lipid composition (bacterial membranes contain a large number of anionic lipids such as phosphatidyl glycerol and cardiolipin and lack cholesterol in their membranes). Gram-negative bacteria have an additional, outer membrane, and our data suggests that a further level of selectivity is expressed there in that there are Gram-positive bacteria-selective peptides that interact poorly with the outer membrane but (presumably) well with cytoplasmic membranes, whereas we have identified peptides that interact with the outer membrane, but are not bactericidal and thus do not interact with cytoplasmic membranes.

Although specific details may vary depending on the peptide, enough data exist to present a general model for the mechanism of action of cationic peptides against Gram-negative bacteria. This process is described below, and can be summarized as a sequence of events involving interaction with lipopolysac-

From: *Methods in Molecular Biology, Vol. 78: Antibacterial Peptide Protocols*
Edited by: W. M. Shafer, Humana Press Inc., Totowa, NJ

charide (LPS) and self-promoted uptake of the peptides across the outer membrane, electrostatic interaction with the negatively charged head groups of lipids in the cytoplasmic membrane, and insertion into the membrane and channel formation leading to leakage of essential nutrients from the cell.

1.1. Dansyl Polymyxin B Displacement Assay

Cationic peptides *(1,2)*, like other polycationic antibiotics *(3)*, traverse the outer membrane using a process termed self-promoted uptake; in contrast small hydrophilic antibiotics such as β-lactams diffuse through the water-filled channels of porin proteins *(4)*. According to the self-promoted uptake model, compounds that access this pathway initially bind to the divalent-cation-binding sites of LPS. To study this, dansyl polymyxin B displacement assays can be performed using purified LPS or whole cells. Dansyl polymyxin B has been shown to bind to the divalent-cation-binding sites of LPS, resulting in greatly enhanced fluorescence of the dansyl group *(5)*. This property led to the development of the above assay for determining the relative LPS-binding affinities of antibiotics based on their ability to competitively displace dansyl polymyxin B from LPS *(5)*.

1.2. Antiendotoxin Activity

As mentioned above, the initial step in the uptake of cationic peptides across the outer membrane is binding to LPS. This binding, which is specific to the lipid A portion of LPS (i.e., endotoxin), can neutralize the ability of LPS to induce tumor neurosis factor (TNF) in macrophage cell lines.

1.3. Lysozyme Lysis Assay

Since the cationic peptides have an affinity for LPS that is three orders of magnitude higher than the native divalent cations, Mg^{2+} or Ca^{2+} *(1)*, they competitively displace these cations. This causes a distortion of outer membrane structure, that has been visualized in the electron microscope as induction of outer membrane blebs *(1)*, and a consequent permeabilization of the membrane to various probe molecules. One of them, lysozyme, is a 14-kDa basic protein that is unable to penetrate intact outer membranes, but can diffuse across disrupted membranes to exert its ability to enzymatically cleave peptidoglycan leading to cell lysis *(6)*. Because of its large size, one would expect that a significant destabilization of the outer membrane would be required for it to penetrate to its peptidoglycan substrate. Cationic peptides can therefore be tested in a lysozyme lysis assay for their ability to permeabilize the outer membrane of the test bacterium and facilitate the uptake of lysozyme.

1.4. NPN Uptake Assay

A second probe molecule, 1-*N*-phenyl-napthylamine (NPN), is an uncharged, hydrophobic fluorescent probe that has been used to study membrane permeabilization *(1,7)*. NPN fluoresces weakly in an aqueous environment, but strongly in the hydrophobic interior of a membrane. When NPN is mixed with cells, it fluoresces weakly since it is unable to breach the outer membrane permeability barrier (or perhaps is rapidly effluxed from such bacteria; *see* **ref. 7**). Upon outer membrane destabilization in the presence of an energy inhibitor, however, it can partition into the hydrophobic environment of the membrane, where it emits a bright fluorescence. NPN is both smaller and more hydrophobic than lysozyme, which enables it to insert into membranes more easily than lysozyme *(7)*, although many cationic peptides cause major disruptions of the outer membrane even at low concentrations, leading to permeabilization to both probes.

1.5. Synergy with Antibiotics

The ability of cationic peptides to act in synergy with certain classical antibiotics *(8)* can be explained in part by their ability to disrupt outer membrane integrity, promoting the uptake of antibiotics across this barrier. Interestingly, the most potent cationic peptides do not have this "enhancer" activity for most antibiotics presumably since they kill cells at concentrations equal to their permeabilizing concentrations *(2)*. This is analogous to the situation for the polycationic antibiotic polymyxin B, which is not an enhancer, whereas its deacylated derivative PMBN (which interacts weakly with cytoplasmic membranes but strongly with outer membranes) is a potent enhancer of antibiotic activity *(9)*. To measure possible synergy, fractional inhibitory concentration (checkerboard) assays can be performed using the test antibiotics in the presence of sub-MIC concentrations (e.g., one-half or one-fourth MIC values) of the peptide *(10)*.

1.6. Planar Lipid Bilayer

Following uptake across the outer membrane, the peptides rapidly associate with the negatively charged head groups of lipids on the cytoplasmic membrane in a cooperative process. The extent of binding corresponds to the zeta potential of the lipids involved, strongly suggesting that binding is governed by electrostatic interactions *(11,12)*. It is uncertain as to whether at this stage the permeability of the target lipid membrane changes, although it is probable that the cationic peptide undergoes a change in conformation and aggregation state as a result of this interaction. At a threshold concentration of peptides bound to the membrane surface, the peptides are able to insert into the membrane and form channels. Although some peptides can insert spontaneously into membranes with little or no transmembrane potential, it seems likely that the membrane potential of living cells (oriented interior negative) is an important factor in peptide insertion. The process of insertion can also cause a conformational change in the cationic peptide (e.g., from unstructured to α-helical), as for example with melittin ***(13)*** and magainins ***(14,15)***. In many cases the peptides are thought to end up spanning the membrane bilayer ***(13,16,17)*** in multimeric complexes. Other peptides are too short to span the bilayer and presumably must stack to form aggregates to permit transmembrane channel formation ***(15,17,18)***. The net effect of channel formation is to disrupt the integrity of bacterial cytoplasmic membranes. This would have the effect of permitting leakage of ions and small metabolites and destroying the ability of bacteria to maintain a transmembrane proton gradient (proton-motive force) with consequent loss of ability to generate ATP and to transport substrates.

The strongest evidence for channel formation has been in model membrane experiments using the planar lipid bilayer technique. In these experiments, the membrane potential, which is provided as an applied voltage, must be oriented positive on the *cis* side (where the cationic peptides are added) and negative on the *trans* side of the membrane (toward which the cationic peptides tend to move as they enter the membrane). This results in an observable increase in conductance as the peptides enter the membrane and forms channels ***(19–23)***. Reversal of the voltage actually causes peptides to leave the membrane ***(19)***. It should be noted, however, that channel formation may not be the only mechanism for cell lysis. Several studies using model liposomes have suggested that lysis can occur by a nonpore mechanism ***(24,25)*** in which cationic peptides form a "carpet-like" layer on the membrane surface that leads to a severe disruption in the lipid bilayer packing and eventual membrane disintegration. In addition it is known that cationic peptides can stimulate the intrinsic autolytic mechanisms of bacteria.

2. Materials

2.1. Dansyl Polymyxin Displacement Assay

2.1.1. Dansyl Polymyxin Synthesis

1. Polymyxin B sulphate (PxB).
2. 0.1*M* $NaHCO_3$.
3. Dansyl chloride (Sigma D2625).
4. Acetone.
5. 10 m*M* Na_2HPO_4, 0.145*M* NaCl.
6. 10 m*M* NaH_2PO_4, 0.145*M* NaCl.

7. Column, 50 × 2.5 cm packed with Sephadex G25 or G50.
8. *n*-Butanol.
9. 5 m*M* HEPES buffer, pH 7.2. Filter sterilize; do not autoclave.
10. UV lamp or UV lightbox.

2.1.2. Dansyl Polymyxin Quantitation (Dinitrophenylation Assay)

1. Polymyxin B sulfate (PxB).
2. 1% (w/v) $Na_2B_4O_7 \cdot 10\ H_2O$.
3. 100 m*M* 2,4 -dinitro-1-fluorobenzene (=1-fluoro-2,4-dinitrobenzene) in ethanol.
4. 2*N* HCl.
5. *n*-Butanol.

2.1.3. Dansyl Polymyxin Displacement Assay

1. 100 µ*M* dansylated polymyxin B (DPX).
2. 5 m*M* HEPES buffer, pH 7.2 (as in **Subheading 2.1.1.**).
3. LPS 3 µg/mL in above HEPES buffer. Store at 4°C. Solutions are stable for several months (*see* **Notes 1, 2,** and **8**).
4. Displacement compounds for testing: stock solutions of polymyxin B at 1.0 mg/mL, gentamicin at 10 mg/mL, and magnesium chloride at 100 m*M*.
5. 3% (w/v) Triton X-100.
6. Fluorescence spectrophotometer (we use a Perkin Elmer [Norwalk, CT] 650-10S machine with a strip chart recorder attached).

2.2. LPS/Endotoxin Neutralization Assay

1. Dulbecco's modified Eagle medium (DMEM): Filter sterilize through a 0.22-µm filter.
2. Hanks' balanced salt solution: Filter sterilize through a 0.22-µm filter.
3. 1*M* HEPES: Sterilize by autoclaving.
4. Hanks'-HEPES buffer: To 500 mL of Hanks' balanced salt solution, aseptically add 12.5 mL sterile HEPES buffer; store at 4°C.
5. L-glutamine: 29.2 mg/mL dissolved in ddH_2O. Filter sterilize through a 0.22-µm filter and store in 6-mL aliquots in 15-mL sterile tubes.
6. Trypan blue: 4 mg/mL dissolved in sterile PBS. Filter sterilize through a 0.44-µm filter and store in 0.4-mL aliquots at room temperature.
7. MTT (3-[4, 5-Dimethylthiazol-2-yl]-2, 5-diphenyltetrazolium bromide; thiazolyl blue): 5 mg/mL dissolved in DMEM (without phenol red) and stored at 4°C in dark or foil covered bottles (this compound is light-sensitive).
8. Actinomycin D: 40 µg/mL dissolved in absolute ethanol and stored at 4°C (this compound is light-sensitive). For use in the TNF assay, dilute 1:10 into RAW cell media.
9. Penicillin: 10^4 U/mL sterile solution.
10. Streptomycin: 1 mg/mL sterile solution.
11. β-mercaptoethanol: 0.1*M* sterile solution.

12. Heat-inactivated fetal bovine serum.
13. RAW cell media: DMEM containing 2.4 m*M* L-glutamine, 60 U/mL penicillin, 6 μg/mL streptomycin, 1.2m*M* β-mercaptoethanol, and 10% heat-inactivated fetal bovine serum (FBS).
14. L929 Growth media: same as RAW cell media, but can substitute heat inactivated horse serum for FBS.
15. TNF assay media: RAW cell media containing 4 μg/mL Actinomycin D.
16. Cell dissociation solution, non enzymatic (Sigma, St. Louis, MO). Store at 4–6°C. Do not freeze.
17. Trypsin-EDTA solution (1X).
18. Test peptides.

2.3. Lysozyme Lysis Assay

1. Luria broth (LB): 10 g tryptone, 5 g yeast extract/ L dH_2O, no added NaCl.
2. 5 m*M* HEPES buffer, pH 7.2 with 5 m*M* sodium azide or 5 m*M* potassium cyanide (to inhibit respiration and prevent active excretion). NaN_3 and KCN are both poisonous. Do not autoclave. Filter sterilize if desired for longer shelf life; sterile buffer is not required for assay.
3. Lysozyme: 5 mg/mL dH_2O.
4. Permeabilizing compounds. Test compounds and positive controls: stock solutions of gentamicin at 10 mg/mL, polymyxin B at 1 mg/mL, and EDTA at 500 m*M*.
5. Spectrophotometer (We use a Perkin Elmer Lambda 3 dual beam spectrophotometer with a strip chart recorder attached).

2.4. NPN Uptake

1. LB broth (*see* **Subheading 2.3.**).
2. For *P. aeruginosa*: 5 m*M* HEPES buffer, pH 7.2, containing 5 m*M* NaN_3 or 5 m*M* KCN (*see* **Subheading 2.3.**). For *E. coli*: 5 m*M* HEPES buffer, pH 7.2, containing 5 m*M* glucose, 5 m*M* NaN_3 or 5 m*M* KCN or 5 μM CCCP (carbonyl cyanide *m*-chlorophenyl hydrazone). CCCP works the best but is unstable and light sensitive. Make stock solutions in ethanol and store at –20°C in the dark (solutions are stable for several years). Add to buffer to make enough working solution for that day.
3. 1-*N*-phenylnapthylamine (NPN). Prepare a 5-m*M* stock and 0.5 m*M* working solutions in acetone. Store at –20°C in foil covered bottles (solutions are stable for several months/1–2 yr).
4. Permeabilizing compounds prepared in serial dilutions at 100X the desired final concentration. Concentrations needed will vary with the compound and the bacteria used, e.g., polymyxin B at 0.2–0.64 mg/mL, gentamicin 1.5–10.0 mg/mL.
5. Fluorescence spectrophotometer with a strip chart recorder attached.

2.5. Fractional Inhibitory Concentration Assay for Synergy

1. Untreated, polypropylene, 96-well, round-bottomed, sterile microtiter trays (Costar 3790, Cambridge, MA) (*see* **Note 23**).

2. Growth media, e.g., Mueller Hinton broth. (*see* **Note 18**).
3. Antibiotic/compound "A," at four times the desired final concentration, in media (*see* **Note 19**).
4. Antibiotic/compound "B," at two times the desired final concentration, in media (*see* **Note 19**).
5. Test strains of bacteria, log-phase or overnight cultures (*see* **Note 20**).
6. Multipipettor(s).

3. Methods

3.1. Dansyl Polymyxin (DPX) Displacement Assay

3.1.1. Dansyl Polymyxin Synthesis (26)

1. Prepare column buffer by adding Na_2HPO_4 and NaH_2PO_4 solutions together in an approx 2:1 ratio until the pH is 7.1. Equilibrate a packed column (50 × 25 cm) with at least 100 mL of buffer.
2. Dissolve 40 mg of polymyxin B in 1.2 mL of 0.1*M* $NaHCO_3$.
3. Dissolve 10 mg dansyl chloride in 0.8 mL acetone. Add to the polymyxin B mixture from above and then incubate in the dark at 23°C for 90 min.
4. Load the mixture on to the column, collect 5- to 6-mL fractions. Monitor fractions using a UV lamp. The dansyl-polymyxin (DPX) fluoresces yellow-orange and comes out as a broad peak (at approx 50–100 mL) before the dansyl-chloride which fluoresces blue-green (at approx 100–130 mL).
5. Pool the DPX fractions. Add one-half the volume of butanol and mix well. Allow the butanol phase to partition or spin gently (100 rpm for 3 min) to separate the phases. Save the butanol phase and place it in shallow glass dish in desiccator. Evacuate the desiccator and place at 37°C until dry, approx 24 h.
6. Dissolve the dried DPX in 3 mL HEPES buffer, and store in aliquots at –20°C, in the dark (lasts for several years).
7. The concentration of the DPX is determined by the dinitrophenylation assay.

3.1.2. Dansyl Polymyxin Quantitation (Dinitrophenylation Assay) (27)

1. Prepare a standard curve using a 1.0 mg/mL stock solution of polymyxin B to give 0, 5, 10, 20, 30, 40, and 50 µg/mL. Use 5 and 50 µL of the DPX solution to be tested. Bring the volume of all samples to 50 µLwith dH_2O.
2. Add 25 µL of 2,4-dinitro-1-fluorobenzene to each sample, mix, and incubate at 37°C for 1 h.
3. Add 1 mL of 2*N* HCl.
4. Add 1 mL of *n*-butanol and vortex.
5. Centrifuge in a clinical centrifuge at 100 rpm for 3 min.
6. Remove the top butanol phase and measure the OD_{420} in a spectrophotometer.
7. Polymyxin B has five free amino groups; DPX has four. Therefore read the corresponding value from the unknown samples off the standard curve and multiply by 1.25 to obtain the quantity of DPX in the sample tube. Calculate the dilution factor to obtain the final concentration of DPX solution.

3.1.3. Dansyl Polymyxin Displacement Assay

1. Set the excitation and emission wavelengths of a fluorescence spectrophotometer to 340 nm and 485 nm respectively, using narrow slit widths of 5 nm.
2. Add 50 µL of 100 µ*M* DPX to 1 mL of 3% Triton X-100. Measure the fluorescence level. Adjust the spectrophotometer sensitivity to give a reading of 90–100% of maximum deflection on the chart recorder.

3.1.3.1. Determining Background Fluorescence of DPX solution

1. Add 5 µL of 100 µ*M* of DPX solution (final amount equals 500 pmol) to 1 mL of HEPES buffer. Measure the fluorescence level.
2. Repeat addition of 5 µL of 100 µ*M* DPX 5–10 times, adding DPX to the same cuvet and measuring the fluorescence for approx 10–20 s after each addition. Each addition should result in a small, equal increase in fluorescence.

3.1.3.2. Determining Saturation of LPS with DPX

1. Add 5 µL of DPX to 1 mL of LPS 3 µg/mL. Measure the fluorescence level.
2. Repeat, adding 5-µL amounts of DPX to the same cuvette and measuring the fluorescence until the level of fluorescence plateaus off and the increase is only a result of the change in background fluorescence.
3. Using the data from the above two steps, determine the amount of DPX that must be added to the 3 µg/mL LPS to give 85–90% of saturation. Call this concentration/amount "Z."
4. Add "Z" amount of DPX to 1 mL of 3 µg/mL LPS. Reset the spectrophotometer so this value gives 90% of maximum scale deflection of the chart recorder (*see* **Note 7**).

3.1.3.3. Determining Inhibition/Displacement of DPX.

1. Add "Z" amount of DPX to 1 mL of 3 µg/mL LPS. Measure the fluorescence level (should be 90%).
2. Add a small aliquot (5–10 µL) of the potential displacer. Measure the fluorescence level for 30–60 s. If the test compound displaces DPX there will be a decrease in fluorescence as the DPX is removed from the LPS and goes into the aqueous solution.
3. Repeat addition of aliquots 5–10 times or until the maximum displacement is reached (i.e., additional compound no longer results in decreased fluorescence) (*see* **Notes 3–6**).
4. Plot the data as percent inhibition versus the concentration of inhibitor (µ*M*). If enough inhibitor is used the maximal % inhibition (I_{max}) can be seen when the fluorescence decrease levels off. The I_{50} is the concentration of compound resulting in half maximal displacement of DPX from LPS. The data can also be plotted as a Lineweaver-Burke plot, plotting reciprocals of each axis of the first plot (i.e., 1/%inhibition vs 1/µ*M* concentration). Extrapolate to determine X and Y intercepts if necessary. I_{max} is then calculated as 100/Y intercept and I_{50} is calculated from –1/X intercept.

3.2. LPS/Endotoxin Neutralization (28) (see Note 9)

3.2.1. Preparation of RAW Cells

1. Grow the murine RAW 264.7 macrophage cell line by seeding 10^6 cells into a 162-cm^2 cell culture flask and incubating at 37°C, 5% CO_2 for 1 wk.
2. Completely remove the RAW cell media and incubate with 10 mL of cell dissociation solution for 10 min at 37°C, 5% CO_2.
3. Remove cells from the flask and dilute with 10 mL RAW cell media.
4. Centrifuge cells at 500*g* for 5 min and resuspend the cell pellet in 5 mL of media.
5. Count the number of cells using a hemocytometer by removing a 0.1-mL aliquot and mixing with 0.4 mL of trypan blue.
6. Dilute the cell suspension to 10^6 cells/mL and add 1 mL of suspension to each well of a 24-well plate.
7. Incubate the plate at 37°C in 5% CO_2 overnight for use in the assay.
8. Save several milliliters of cell suspension from **step 6** to seed a new flask.

3.2.2. Induction of TNF

1. After overnight incubation, the medium is aspirated from each of the wells in the 24-well plate.
2. Wash each well with 1 mL of Hanks' balanced salt solution and aspirate off.
3. Either synthetic LPS or whole bacteria can be used to induce TNF production. If synthetic LPS is used, add 0.1 mL to each well for a final concentration of 100 ng/mL, and continue with **step 6** (*see* **Note 10**).
4. For experiments using whole bacteria, overnight cultures are diluted to an OD_{600} of 0.3 (10^8 cells/mL).
5. Further dilute the bacteria 1:10, and perform a viable count measurement to determine the exact number of bacteria used.
6. Add 0.1 mL of the diluted bacteria (approx 10^6 cells) to a Millipore (Bedford, MA) transwell filter insert (<0.2 μm) and place in the wells.
7. Add peptide to specific wells at the desired concentration.
8. Add RAW cell media (without penicillin/streptomycin) to all of the wells to give a final volume of 1.0 mL, and incubate the plate for 6 h at 37°C and 5% CO_2.
9. Remove supernatants from the wells and store at 4°C overnight.

3.2.3. Preparation of L929 Cells

1. Grow the TNF-sensitive mouse fibroblast cell line by seeding 10^6 cells into a 162-cm^2 cell culture flask and incubating at 37°C, 5% CO_2 for 1 wk.
2. Completely remove the L929 cell media and incubate with 10 mL of trypsin EDTA solution for 5–10 min at 37°C, 5% CO_2.
3. Remove cells from the flask and dilute with 10 mL L929 cell media.
4. Centrifuge cells at 500*g* for 5 min and resuspend the cell pellet in 5 mL of media.
5. Count the number of cells using a hemocytometer by removing a 0.1-mL aliquot and mixing with 0.4 mL of trypan blue.

6. Dilute the cell suspension to 10^6 cells/mL for use in L929 cell toxicity assay.
7. Save several milliliters of cell suspension from **step 6** to seed a new flask.

3.2.4. Cell Cytotoxicity Assay

The levels of TNF released into the supernatants can be measured using the following L929 cell cytotoxicity assay.

1. Add 50 µL of TNF media to all the wells of a 96-well microtiter plate except to those in the first row.
2. Add 10 µL of murine TNF standard (20 ng/mL) and 90 µL of TNF medium in duplicate to wells A1 and A2 in the first row and dilute 1:2 (i.e., –50 µL from A1 into B1, 50 µL from B1 into C1, and so on) down the plate until the second to last row. Discard the extra 50 µL from row G; do not add to row H (*see* **Note 11**).
3. 75 µL of the test samples comprising the supernatants from the RAW cell assay are then added in duplicate to the next columns in the first row (i.e., first sample added to A3 and A4, second sample to A5 and A6, and so on). The samples are diluted 1:3 down the plate until the second to last row (i.e., 25 µL from A3 into B3, 25 µL from B3 into C3, and so on). Discard the extra 25 µL from row G; do not add to row H.
4. All of the wells should now contain a volume of 50 µL.
5. Add 100 µL of the previously prepared L929 cell suspension (10^6 cells/mL) to each of the wells except H11 and H12.
6. The last row (H1–H10) contains only TNF media and cells and is used as a negative control; wells H11 and H12 contain only TNF media and are used as blank control.
7. Incubate the plate at 37°C, 5% CO_2 for 2 d.

3.2.5. TNF Measurement

1. Aspirate the medium from the plate and replace with 0.1 mL of the dye MTT (0.5 mg/mL) in DMEM without phenol red.
2. Incubate the plate at 37°C and 5% CO_2 for 3 h.
3. Remove the dye and replace with 0.1 mL absolute ethanol.
4. Incubate the plate at room temperature for 10–15 min to dissolve the formazan dye crystals.
5. Read plate at 570 nm in an ELISA plate reader with a 690-nm reference filter. One unit of TNF activity is defined as the amount of TNF required to kill 50% of the L929 cells.

3.3. Lysozyme Lysis Assay

1. Use 1 mL overnight culture to inoculate 50 mL LB. Grow cells to mid-log phase OD_{600} 0.4–0.6 (*see* **Note 12**).
2. Centrifuge at 3000*g* for 10 min at 23°C. Resuspend in HEPES buffer, wash once, and then resuspend to a final OD_{600} of 0.5 (*see* **Note 13**).

3. Add 1 mL of cells to the cuvette. Measure OD_{600} for approx 10 s.
4. Add lysozyme to a final concentration of 50 µg/mL. Measure OD for approx 10 s. There should be no lysis.
5. Add permeabilizer to the desired final concentration. Add small volumes (5–10 µL) of stock solutions so cells/lysozyme are not heavily diluted. Measure OD for at least 1 min.
6. Measure percent cell lysis as the percent decrease in OD_{600} at a set time point after addition of the permeabilizer, i.e. 3 min. If a chart recorder is attached, compare lysis curves. The result depends on the concentration of permeabilizer and a plot of percent lysis vs permeabilizer concentration can be developed.
7. Measure negative controls: cells and lysozyme (no permeabilizer), buffer and lysozyme and permeabilizer (no cells), cells and permeabilizer (no lysozyme). Note: some peptides stimulate autolysis (*see* **Notes 14** and **15**).

3.4. NPN Uptake Assay

1. Use 1 mL overnight culture to inoculate 50 mLs of LB. Grow cells to mid-log phase (OD_{600} = 0.4–0.6) (*see* **Note 16**).
2. Centrifuge at 3000*g* for 10 min at 23°C. Resuspend in HEPES buffer, wash once, and then resuspend to a final OD_{600} of 0.5. Maintain cells at 23°C.
3. Set the excitation and the emission wavelengths of a fluorescence spectrophotometer to 350 nm and 420 nm, respectively, using narrow slit widths of 5 nm. Fluorescence is measured in arbitrary units. To standardize experiments, choose one set of conditions and adjust the sensitivity of the spectrophotometer and the chart recorder to a set value each time. For example to 1 mL of cells (OD_{600} = 0.5) add 10 µm NPN and 6.4 µg/mL polymyxin B and adjust sensitivity to give 90% of maximum scale deflection of the chart recorder. Under these conditions the NPN background level (without polymyxin B) is usually 5–15%.
4. Add 1 mL of cells to the cuvet and measure the fluorescence level after 5 s.
5. Add 10 µ*M* NPN to the same cuvet. Mix. Measure the new fluorescence level after 5 s.
6. Add 5–10 µL of permeabilizer to the same cuvet. Mix. Measure the fluorescence level until there is no further increase (*see* **Note 17**).
7. Using a fresh cuvet of cells with 10 µ*M* NPN each time, repeat steps 4–6 over a wide range of concentrations for each compound to give the full range of levels of fluorescence. Dilute the compounds so that equal volumes are added to each cuvette of cells, rather than having one concentration of compound and adding different volumes.
8. Measure as negative controls: cells + NPN + antibiotic "solvent," NPN in buffer only (no cells, no antibiotics control), NPN in buffer only + antibiotic (no cells control).
9. To plot data, subtract the background level of fluorescence of the NPN only from the total fluorescence and plot versus the concentration of the compound.

3.5. Fractional Inhibitory Concentration Assay for Synergy (29)

1. Pipet 100 μL of media only into each well of the 96-well microtiter tray in which columns are numbered 1–12 and rows A–H.
2. Add 100 μL of compound A (4X) to each well in column 1.
3. Do doubling dilutions *across* the plate from columns 1–11: using a multipipettor, carefully mix all the wells in column 1 , then transfer 100 μL from that column to the next one. Repeat until column 11, then discard the last 100 μL; do not add it to column 12.
4. Add 100 μL of compound B (2X) to each well in the top row A of the plate.
5. Do doubling dilutions *down* the plate from rows A–G: using a multipipetor, carefully mix all the wells in row A , then transfer 100 μL from that row to the next one. Repeat until row G, then discard the last 100 μL; do not add it to row H.
6. Add 100 μL of media only to row H. Mix, then discard 100 μL. This will bring the concentration of compound A in each column of row H to the same concentration as the other rows for each respective column.
7. Add 5–10 μL of bacteria to each well to give a final inocculum of 5×10^5CFU/mL.
8. Incubate at 37°C and read visually after 18–24 h. Well H12 serves as a positive control as it contains only the growth medium and the bacteria.
9. Calculate the MIC for each compound from row H (for compound A) and column 12 (for compound B). The MIC is read as the minimal concentration necessary to inhibit growth by at least 50%. Calculate the FIC index.

3.5.1. Calculation of the FIC Index

The FIC index is calculated from the formula:

$$\text{FIC index} = \text{FIC A} + \text{FIC B} = (\text{A})/(\text{MIC A}) + (\text{B})/(\text{MIC B})$$

where (A) is the concentration of compound A in the microtiter well that is the lowest inhibitory concentration of compound A in its row. (MIC A) is the MIC of drug A alone. FIC A is the fractional inhibitory concentration of compound A. (B), (MIC B), and FIC B are defined for compound B in the same way as for compound A.

An FIC index of ≤0.5 indicates synergy, when the results with two drugs are significantly greater than the additive response. An FIC index of >1.0 indicates antagonism, when the results with two drugs are significantly less than the additive response. An FIC index = 1.0 indicates additivity; when the results with two drugs are equal to the sum of the results for each of the drugs used separately. An FIC index = FIC A or FIC B alone indicates autonomy, when the results with two drugs do not significantly differ from the result with the most effective drug alone.

In the example in **Fig. 1**, at combination * (well E6), (A) = 2, (B) = 1.25; therefore, the calculation is as follows:

	row	1	2	3	4	5	6	7	8	9	10	11	12
column	conc	64	32	16	8	4	2	1	.5	.25	.12	.06	.03
A	20												
B	10												
C	5											+	+
D	2.5											+	+
E	1.25						*	+	+	+	+	+	+
F	0.63						+	+	+	+	+	+	+
G	0.32					+	+	+	+	+	+	+	+
H	0.15				+	+	+	+	+	+	+	+	+

Fig. 1. Example to calculate FIC index. Growth is indicated by +, the MIC of compound A = 16 (well H3), the MIC of compound B = 10 (well B12). *Indicates well used to illustrate calculation of FIC index.

$$\text{FIC index} = (\text{A})/(\text{MIC A}) + (\text{B})/(\text{MIC B})$$

$$\text{FIC index} = 2/16 + 1.25/10 = 0.125 + 0.125 = 0.25$$

Since 0.25 is ≤0.5, this therefore indicates synergy between compounds A and B. Please note that the FIC index for well F5 is equal to 0.31 and also indicates synergy. Usually the lowest calculation of FIC index is used (*see* **Notes 21** and **22**).

3.6. Planar Lipid Bilayer

The planar lipid bilayer method offers a rapid way to experimentally measure the ability of peptides to form transbilayer channels. In general, cationic peptides form multistate channels, and planar bilayer experiments demonstrate a substantial range of channel sizes, with single channel conductances (which reflect size) varying from 10–2000 pS (**refs. *19–23***) and lifetimes ranging from milliseconds to seconds. The planar lipid bilayer method, described briefly below, utilizes specialized apparatus that usually requires some training in its use.

The central part of the apparatus comprises a chamber that is machined from a 5 × 2.5 × 3-cm block to create two equal compartments, separated by a 1-mm Teflon divider. One of these compartments contains a viewing window, and the Teflon divider is perforated by a 0.1- to 2-mm^2 hole. The hole is anointed at its edges with a lipid solution and dried under a jet of hot air to provide a surface to which a membrane can adhere. The compartments are then each filled with 6 mL of a salt solution (e.g., 1*M* KCl). The hole is then covered with lipid by wiping with a Teflon rod onto which 5 µL of the lipid solution has been pipetted. Successful "painting" of the lipid over the hole is assessed by measuring a high resistance when a voltage is passed across two electrodes dipping into the two compartments. Within a short time, the lipid thins out until it forms a bilayer. This can be observed using a short focal length telescope in incident light coming from a suitable focused light source such as a microscope light, since the observed lipid changes from multicolored to black, due to the optical properties of lipid globules and lipid bilayers respectively. This gives the methodology its alternate name black lipid bilayers.

To the electrode dipping into the solution in one compartment, a direct current voltage source is attached. To the other electrode is attached a current amplifier (Keithly 427 [Cleveland, OH]), an oscilloscope to monitor the amplified signal (Tektronix 511A [Beaverton, OR]) and a rapid response chart recorder (Houston Instruments 4512). With the naked membrane, application of a voltage (typically 30–100 mV) results in a very small current (approx 2 pA) since bilayers have little permeability to ions. When peptide is added to the compartment containing the cathode (*cis* side) on one side of the membrane, depending on its affinity, it will incorporate into the membrane at a threshold voltage and form a conduit for the movement of ions through the channel. This can be observed, after signal amplification, as a stepwise increase in conductance in both real time on the chart recorder and at a faster resolution on the oscilloscope.

Some of the channel properties that can be easily checked by varying experimental parameters are whether the channel is water-filled, whether there is a strong selectivity for cations over anions, or vice versa, the influence of lipid composition, whether the channel aggregates in the membrane, the effect of voltage on channel properties and on the ability of channels to insert into the membrane (voltage induction) or open under a voltage (voltage gating), whether the channel permits only unidirectional flux of ions, and the variability in the sizes of individual channels. The major readout, however, is average single channel conductance of the channel in given salts, which is itself proportional to the volume of the channel and its geometry.

With a slightly different setup in which only a voltage source and a multimeter (Keithly 610) are connected to the electrodes, one can measure macroscopic conductance, and determine such properties as voltage dependence and selectivity for one ion over another.

4. Notes

1. LPS may be purchased or prepared from bacterial cells ***(8)***.
2. LPS is more stable and gives more consistent results after further purification by chloroform/methanol extraction to remove contaminating lipids, and conversion to the sodium salt form which exchanges sodium for magnesium at cross-bridging sites (***30***, but omit $MgCl_2$ dialysis step).
3. Try to use a concentration of displacer that will give several relatively even step decreases in fluorescence of 5–15% before reaching maximum displacement.

4. Usually the decrease in fluorescence is immediate and fairly stable. If necessary measure the fluorescence decrease for a longer time until the decrease levels off.
5. If the cuvet containing LPS and DPX alone gives erratic or drifting fluorescence recordings, try briefly sonicating the LPS solution.
6. Many common antibiotic solvents such as alcohols, and acidic or basic solutions will cause decreases in fluorescence (i.e., false positive results). Similarly, detergents can cause increases in fluorescence. Be sure to include a negative control using your "solvent" only.
7. Once the "Z" concentration of DPX is determined, the background and saturation points do not have to be retested if using the same conditions (DPX concentration and LPS stock). For further displacement experiments, just begin at **Subheading 3.1.3.3.**
8. Lipid A at 4 μg/mL or whole cells may be used instead of LPS. For whole cells, grow up cells to mid-log phase in LB. Pellet, wash in HEPES buffer containing 5 m*M* sodium azide (to inhibit respiration). Resuspend to an OD_{600} of 0.5.
9. Plan the experiment properly so that everything will be ready at the necessary time. On d 1, seed the RAW cells (**Subheading 3.2.1., step 1**); on d 3, seed the L929 cells (**Subheading 3.2.3., step 1**). On d 8 set up the RAW cells into the 24-well plates (**Subheading 3.2.1., steps 2–7**), then induce them with TNF on d 9 and save the supernatants (**Subheading 3.2.2.**). On d 10, use the supernatants to set up the microtiter plates for the L929 cell toxicity assay (**Subheading 3.2.4.**); do all the dilutions before getting the L929 cell ready (**Subheading 3.2.3., steps 2–8**). Incubate the plates and then read on d 12 (**Subheading 3.2.5.**).
10. If whole bacteria are used in the induction of TNF (**Subheading 3.2.2., steps 2–5**), they can be added without using the Millipore transwell filters, but after removing supernatants centrifuge them in 0.2-μm filter Eppendorf tubes and store the clarified supernatants at 4°C overnight.
11. The TNF standard (positive control) is only required on one plate for the L929 cell toxicity assay (**Subheading 3.2.4., step 2**).
12. Other growth media may be used, but high levels of cationic compounds such as Mg^{2+} and Ca^{2+} may interfere by binding to the same binding sites that the permeabilizer would access.
13. Spin and keep cells at 23°C to avoid "leaky" cells. Do not keep on ice! *Pseudomonas aeruginosa* is especially affected, and can become very leaky even if cells are only refrigerated during centrifugation. Resuspend cell pellets gently with a pipet; do not vortex.
14. If the test compound causes cell lysis without lysozyme addition, try to stabilize cell osmolarity by using 80 m*M* (or higher) NaCl in the buffer. Lysis by the test compound alone can also be subtracted as background.
15. Watch the cuvets for any signs of cell clumping, which will also cause changes in OD. If in doubt, check the cells using a microscope.
16. Cell stability is very important in this assay. See notes in **Notes 12** and **13** regarding the growth of cells. Measure the background level of fluorescence of the cells alone plus NPN for at least 1 min at the beginning of the assays. Keep an eye on

the background levels throughout successive experiments and again measure the value with the cells alone plus NPN if the background seems to have increased substantially, or at least every hour to ensure cells are stable. Do not continue the experiment once the cells have become unstable (after 1–3 h). If the cells seem unstable from the very beginning, it is better to discard them and grow another batch.

17. Add the permeabilizer soon after adding the NPN, since the level of fluorescence with the permeabilizer is usually somewhat lower if it is, for example, added 1 min as compared to adding it 10 s after NPN addition.
18. Different growth media may be used but several factors can influence MIC results. Mueller Hinton is the most commonly used and standardized media for MIC measurements.
19. Use concentrations of antibiotic ranging from at least twofold above the MIC to at least fourfold below the MIC.
20. Generally, overnight cultures are around 10^9–10^{10} CFU/mL, therefore dilute 1/10,000 and use 5–10 μL to inoculate, giving a final inoculum of approx 10^5 CFU/mL. Check inoculum by plate counts. Adjust dilutions for further experiments as necessary for your strains. Try to ensure final inoculum is between 10^4–10^6 CFU/mL for accurate MIC values. The MIC results for peptides are not usually different using log-phase cells or overnight cultures, but many antibiotics target growing cells and require log-phase cells for accurate MIC values.
21. Some compounds do not inhibit growth (i.e., have no MIC themselves), but will be synergistic with other compounds. These compounds usually permeabilize the cells, allowing more of the other compound (that inhibits the bacteria) into the cell. A true FIC index cannot be calculated, since there is no MIC, but the pattern of synergy can be easily seen.
22. When calculating FIC indexes, each well will not usually have the same result. Two compounds may be synergistic at one particular combination but not at others.
23. Do not use tissue culture or ELISA plates; they are treated and tend to bind cationic peptides. Polypropylene also binds peptides less than standard polystyrene plates.

Acknowledgments

The financial assistance of the Canadian Cystic Fibrosis Foundation, the Medical Research Council of Canada, the Canadian Bacterial Diseases Network, and Micrologix Biotech is gratefully acknowledged.

References

1. Sawyer, J. G., Martin, N. L., and Hancock, R. E. W. (1988) Interaction of macrophage cationic proteins with the outer membrane of *Pseudomonas aeruginosa. Infec. Immun.* **56,** 693–698.
2. Piers, K. L. and Hancock, R. E. W. (1994) The interaction of a recombinant cecropin/melittin hybrid peptide with the outer membrane of *Pseudomonas aeruginosa. Molec. Microbiol.* **12,** 951–958.

3. Hancock, R. E. W. (1981) Aminoglycoside uptake and mode of action—with special reference to streptomycin and gentamicin. II. Effects of aminoglycosides on cells. *Antimicrob. Chemother.* **8,** 429–445.
4. Hancock, R. E. W. (1991) Bacterial outer membranes: evolving concepts. *ASM News.* **57,** 175–182.
5. Moore, R. A., Bates, N. C., and Hancock, R. E. W. (1986) Interaction of polycationic antibiotics with *Pseudomonas aeruginosa* lipopolysaccharide and lipid A studied by using dansyl-polymyxin. *Antimicrob. Agents Chemother.* **29,** 496–500.
6. Hancock, R. E. W and Wong, P. G. W. (1984) Compounds which increase the permeability of the *Pseudomonas aeruginosa* outer membrane. *Antimicrob. Agents Chemother.* **26,** 48–52.
7. Loh, B., Grant, C., and Hancock, R. E. W. (1984) Use of the fluorescent probe 1-N-phenylnaphthylamine to study the interactions of aminoglycoside antibiotics with the outer membrane of *Pseudomonas aeruginosa. Antimicrob. Agents Chemother.* **26,** 546–551.
8. Darveau, R. and Hancock, R.E.W. (1983) Procedure for isolation of bacterial lipopolysaccharides from both smooth and rough *Pseudomonas aeruginosa* and *Salmonella typhimurium* strains. *J. Bacteriol.* **155,** 831–838.
9. Vaara, M. (1992) Agents that increase the permeability of the outer membrane. *Microbiol Rev.* **56,** 395–411.
10. Piers, K. L., Brown, M. H., and Hancock, R. E. W. (1994) Improvement of outer membrane-permeabilizing and lipopolysaccharide-binding activities of an antimicrobial cationic peptide by C-terminal modification. *Antimicrob. Agents Chemother.* **38,** 2311–2316.
11. Matsuzaki, K., Harada M., Funakoshi, S., Fujii, N., and Miyajima, K. (1991) Physiochemical determinants for the interactions of magainins 1 and 2 with acidic lipid bilayers. *Biochem Biophys Acta.* **1063,** 162–170.
12. Sekharam, K. M., Bradrick, T. D., and Georghiou, S. (1991) Kinetics of melittin binding to phospholipid small unilamellar vesicles. *Biochem Biophys Acta.* **1063,** 171–174.
13. Vogel, H. and Jahnig, F. (1986) The structure of melittin in membranes. *Biophys J.* **50,** 573–582.
14. Bechinger, B., Zasloff, M., and Opella, S. J. (1992) Structure and interactions of magainin antibiotic peptides in lipid bilayers: a solid-state nuclear magnetic resonance investigation. *Biophys J.* **62,** 12–14.
15. Williams, R. W., Starman, R., Taylor, K. M. P., Gable, K., Beeler, T., Zasloff, M., and Covell, D. (1990) Raman spectroscopy of synthetic antimicrobial frog peptides magainin 2a and PGLa. *Biochemistry* **29,** 4490–4496.
16. Sipos, D., Andersson, M., and Ehrenberg, A. (1992) The structure of the mammalian antibacterial peptide cecropin P1 in solution, determined by proton-NMR. *Eur. J. Biochem.* **209,** 163–169.
17. Andreu, D., Ubach, J., Boman, A., Wahlur, B., Wade, D., Merrifield, R. B., and Boman, H. G. (1992) Shortened cecropin A-melittin hybrids. Significant size reductions retains potent antibiotic activity. *FEBS Letts.* **296,** 190–194.

18. Agawa, Y., Lee, S., Ono, S., Aoyagi, H., Ohno, M., Tamguchi, T., Anzai, K., and Kirino, Y. (1991) Interaction with phospholipid bilayers, ion channel formation, and antimicrobial activity of basic amphiphilic alpha-helical model peptides of various chain lengths. *J. Biol. Chem.* **266,** 20,218–20,222.
19. Christensen, B., Fink, J., Merrifield, R. B., and Mauzerall, D. (1988) Channel forming properties of cecropins and related model compounds incorporated into planar lipid membranes. *PNAS.* **85,** 5072–5076.
20. Kagan, B. L., Selsted, M. E., Ganz, T., and Lehrer, R. I. (1990) Antimicrobial defensin peptides form voltage-dependent ion-permeable channels in planar lipid bilayer membranes. *PNAS.* **87,** 210–214.
21. Hanke, W., Methfessel, C., Wilmsen, H. U., Katz, E., Jung, G., and Bohem, G. (1983) Melittin and a chemically modified trichotoxin form alamethicin-type multistate pores. *Biochem. Biophys. Acta* **727,** 108–114.
22. Kordel, M., Benz, R., and Sahl, H. G. (1988) Mode of action of the staphylococcin like peptide Pep5: voltage dependent depolarization of bacterial and artificial membranes. *J. Bacteriol.* **170,** 84–88.
23. Cociancich, S., Ghazi, A., Hetru, C., Hoffman, J. A., and Letellier, L. J. (1993) Insect defensin, an inducible antibacterial peptide, forms voltage-dependent channels in *Micrococcus luteus. Biol. Chem.* **268,** 19,239–19,245.
24. Pouny, Y., Rapaport, D., Mor, A., Nicolas, P., and Shai, Y. (1992) Interaction of antimicrobial dermaseptin and its fluorescently labeled analogues with phospholipid membranes. *Biochemistry* **31,** 12,416–12,423.
25. Gazit, E., Boman, A., Boman, H. G., and Shai, Y. (1995) Interaction of the mammalian antibacterial peptide cecropin P1 with phospholipid vesicles. *Biochemistry* **34,** 11,479–11,488.
26. Schindler, P. R. G. and Tueber, M (1975) Action of Polymyxin B on bacterial membranes: morphological changes in the cytoplasm and in the outer membrane of *Salmonella typhimurium* and *Escherichia coli* B. *Antimicrob. Agents Chem.* **8,** 95–104.
27. Bader, J. and Teuber, M. (1973) Binding to the O-antigenic lipopolysaccharide of *Salmonella typhimurium. Z. Naturforsch.* **28c,** 422–430.
28. Kelly, N. M., Young, Y., and Cross, A. S. (1991) Differential induction of tumor necrosis factor by bacteria expressing rough and smooth lipopolysaccharide phenotypes. *Infect. Immun.* **59,** 4491–4496.
29. Amsterdam, D. (1991) Antimicrobial combinations, in *Antibiotics in Laboratory Medicine.* (Lorian, V., ed.) Williams and Wilkins, Baltimore, pp. 432–492.
30. Peterson, A. A., Hancock, R. E. W., and McGroarty, J. (1985) Binding of polycationic antibiotics and polyamines to lipopolysaccharides of *Pseudomonas aeruginosa. J. Bacteriol.* **164,** 1256–1261.

14

The Genetic Basis of Microbial Resistance to Antimicrobial Peptides

Eduardo A. Groisman and Arden Aspedon

1. Introduction

Small cationic peptides with antibiotic properties have been isolated from a diverse array of evolutionarily divergent organisms, including insects, amphibians, mammals, and plants. They contribute to the innate immunity of the host by fending off opportunistic (i.e., environmental) microorganisms. Moreover, antimicrobial peptides present a chemical barrier early in infection before the mammalian host induces the specific type of immune response constituted by antibodies and T cells *(1,2)*. Microorganisms have coexisted with their animal hosts for millions of years and have, in turn, evolved strategies that enable them to avoid or withstand the various microbicidal activities of the host *(3)*. For example, the mammalian pathogen *Salmonella typhimurium* has several genes that confer resistance to host defense peptides, presumably allowing it to successfully colonize host tissues that are rich in antimicrobial peptides. The demonstration that mutants of *S. typhimurium* that are hypersusceptible to the killing effects of host-defense peptides are attenuated for virulence in mice has established that resistance to small cationic peptides is a virulence property of *Salmonella* *(4)* and, potentially, of other enteric pathogens.

Although antimicrobial peptides differ in length and primary amino acid sequence, they are usually short and positively charged at neutral pH. This suggests that there is an initial electrostatic interaction between the positively charged peptide and the negatively charged surface of the bacterial cell. Many antimicrobial peptides form voltage-gated channels in lipid bilayers *(5–7)*, suggesting that they kill bacteria by depolarizing the cytoplasmic membrane. Indeed, insect defensins have been shown to cause leakage of essential intracellular contents *(8)*, whereas

From: *Methods in Molecular Biology, Vol. 78: Antibacterial Peptide Protocols*
Edited by: W. M. Shafer, Humana Press Inc., Totowa, NJ

treatments that decrease the bacterial membrane potential can often prevent cell killing. Furthermore, defensins can permeabilize the outer membrane of Gram-negative bacteria, but killing only follows inner membrane permeabilization ***(9)***. Thus, determinants conferring resistance to host defense peptides can be classified into three broad categories: those that interfere with the ability of a peptide to reach its target; those that destroy or inhibit the biological activity of an antibiotic peptide; and those that are necessary to restore the physiological balance that was perturbed as a consequence of peptide action. In addition to these categories, there is a fourth group of resistance determinants constituted by the regulatory proteins that control expression of the determinants belonging into each to the three categories listed above.

1.1. General considerations

Several experimental strategies have been adopted to identify bacterial determinants that confer resistance to host defense peptides. Many of these experiments have been carried out with the Gram-negative bacterium *S. typhimurium* because of the ease with which it can be genetically manipulated. These experiments have revealed several important conclusions that are applicable to other microbial species as well.

First, resistance to antimicrobial peptides is multifactorial. Several resistance loci have been identified in a single bacterial species. Moreover, the analysis of mutants and multicopy number suppressors with altered peptide susceptibility has indicated that a given determinant can confer resistance to either a single antimicrobial peptide or to various, structurally unrelated, peptide antibiotics. Second, peptide resistance is transcriptionally regulated. Two different regulatory systems, PhoP/PhoQ and PmrA/PmrB, control resistance to host defense peptides. These regulatory systems belong to a protein family in which transcription is modulated in response to specific environmental signals. For example, the PhoP/PhoQ system modifies transcription of PhoP-regulated genes in response to the concentration of environmental Mg^{2+} ***(10)***. When wild-type *Salmonella* is grown in media containing micromolar levels of Mg^{2+}, PhoP-activated genes are expressed and the microbe displays resistance to the frog-derived peptide magainin 2 ***(10)***. On the other hand, growth in millimolar concentrations of Mg^{2+} represses transcription of PhoP-activated genes and renders *Salmonella* susceptible to magainin 2 (susceptibility approaches that exhibited by a *phoP* mutant). Third, the media used to grow the microbe previous to exposure to the antimicrobial peptide (which may be similar or different from the buffer used during the assay) influences the outcome of the assay. As stated above, the levels of Mg^{2+} in the growth media dramatically alters the susceptibility of *Salmonella* to magainin 2. Likewise, exposure to sublethal acid pH increases polymyxin resistance in *Salmonella* ***(11)***.

In sum, the phenotypic display of susceptibility or resistance by a given microbe only applies to the particular growth and assay conditions used. During stress, microbes often synthesize proteins that mediate the adaptation not only against the particular stress condition tested but also against other, unrelated, stresses *(12)*. When testing for susceptibility/resistance to antimicrobial compounds microorganisms are typically grown in rich media to achieve optimal and rapid growth. However, peptide resistance determinants may not be synthesized under these conditions resulting in a phenotypic display of peptide susceptibility when the microbe is actually resistant in vivo. Furthermore, it is important to keep in mind that whereas one may be able to identify a buffer that mimics the relevant tissue in which a peptide normally functions, it is more difficult to guess the growth condition that induces expression of resistance determinants or the kinetics with which they are synthesized.

1.2. Genetic Strategies to Identify Peptide Resistance Determinants

A combination of classical and molecular genetic approaches have been used to identify genes that confer resistance to host defense peptides. The majority of the examples discussed in this chapter are based on experiments carried out with *S. typhimurium*, which, being amenable to the most sophisticated techniques of genetic analysis, has become a model organism to examine the interaction between microbes and host-defense peptides.

Searches for resistance determinants have been based in one of the following experimental strategies:

1. The isolation of mutants that are more susceptible to a given peptide than the wild-type strain.
2. The isolation of mutants that are more resistant to a particular peptide than the wild-type parent.
3. The identification of genes that, when present in multicopy number plasmids, confer increased peptide resistance to a wild-type strain or to a peptide hypersusceptible mutant.

When attempting large genetic screenings to identify determinants involved in peptide resistance, there is a very important factor that should be taken into consideration: antimicrobial peptides are expensive (either in the actual cost of a synthetic peptide or in the effort that requires its purification). Thus, economic considerations may limit the number of mutants that can be screened or force the use of related compounds that may mimic the activity of the peptide of interest.

1.2.1. Isolation of Peptide-Sensitive Mutants

Transposon mutagenesis provides the most efficient way to isolate loss-of-function mutations that result in peptide susceptibility. The advantage of transposon mutagenesis over chemical or radiation mutagenesis is that the mutated gene is physically marked with an antibiotic resistance marker. This is important because one can then easily establish that the mutant strain harbors a single (transposon-induced) mutation that is responsible for the phenotype of interest. Moreover, the use of transposons facilitates mapping of a resistance locus and cloning of a mutant copy of the gene of interest. The main disadvantage that the use of transposons presents is that it is practically impossible to recover transposon insertions in essential genes, some of which could be involved in peptide resistance. Readers interested in the various genetic approaches to the isolation and characterization of mutants by the use of transposon or chemical (*see below*) mutagenesis are referred to the several reviews that comprise vol. 204 of *Methods in Enzymology,* which is entirely devoted to Bacterial Genetic Systems ***(13)***.

To identify genes required for resistance to small antimicrobial peptides, our laboratory constructed a bank of 20,000 mutants of a peptide-resistant strain of *S. typhimurium* using the transposon MudJ, and screened individual mutants for susceptibility to protamine, a 32-amino acid cationic peptide from salmon sperm. Protamine was chosen as the prototype antimicrobial compound because we had previously established that it preferentially kills *phoP Salmonella* (a mutant that displays hypersensitivity to several antimicrobial peptides including cecropins, magainins and defensins) and, unlike other antimicrobial peptides, it is inexpensive and easily available for such a large-scale genetic screening.

We used Southern hybridization analysis of mutant DNA probed with a MudJ-derived fragment to demonstrate that there was a single MudJ insertion per haploid genome, and that the MudJ transposon had inserted in different sites of the genome for the 12 hypersensitive mutants that were analyzed. Furthemore, we carried out phage transduction experiments to establish that peptide susceptibility was due to the insertion of the MudJ in a peptide resistance locus. Further molecular and genetic experiments allow us to defined eight distinct resistance loci among the 12 investigated mutants.

1.2.2. Isolation of Peptide-Resistant Mutants

The use of spontaneous or chemically-induced mutations may be the method of choice to look for peptide-resistant mutants. Single amino acid substitutions in resistance determinants may give rise to proteins with new activities. For example, Mäkela and coworkers and Roland and coworkers independently isolated mutants of *S. typhimurium* that enhance resistance to polymyxin B ***(14,15)***. The mutations were mapped to the *pmrA* locus, the molecular characterization of which demonstrated that it encodes a transcriptional regulatory protein of the two-component family ***(15)***. The mutant PmrA protein has an Arg for His change that presumably results in a strain with higher levels of expression of PmrA-regulated genes.

Resistant mutants can also be isolated by enhancing expression of resistance genes through the use of transposons that carry a promoter that reads off the end of the transposon and into the chromosomal DNA at the site of insertion. Some of these transposons have regulatable promoters and others are constitutively expressed. Finally, for those peptides that appear to interact with a protein receptor in the bacterium (e.g., apidaecin from honey bees), it should be possible to isolate resistant mutants by inactivating the receptor with a transposon as described in the previous section.

1.2.3. Identification of Genes that Confer Peptide-Resistance when in High Copy

Resistance genes can also be identified by their ability to confer peptide resistance when present in multicopy-number plasmid vectors. The advantage of this method is that the isolation of the resistance clone results in the simultaneous cloning of the gene of interest. After a particular plasmid is isolated, subcloning and mutagenesis experiments are performed to localize the resistant determinant within the plasmid. To establish whether the identified gene normally participates in peptide resistance, one would have to generate a mutation in the plasmid copy of the gene, transfer the mutation back to the chromosome by reverse genetics, and then examine the peptide resistance phenotype of the resulting mutant strain.

Roland and coworkers recovered a gene, *pmrD*, that, when present in a multicopy number plasmid, conferred resistance to polymyxin ***(16)***. The *pmrD* gene encodes an 85-amino acid polypeptide that displays weak similarity to a protease from Rous sarcoma virus. Disruption of the chromosomal copy of the *pmrD* gene resulted in a strain that displayed wildtype levels of polymyxin resistance. Our laboratory has isolated multicopy number suppressors that restore protamine resistance to a *phoP* mutant of *S. typhimurium*. These clones, which were originally isolated on LB agar protamine (1 mg/mL) plates, conferred resistance to different levels of protamine as determined by the use of gradient protamine plates (*see* **Subheading 3.**). The *phoP* gene encodes a transcription factor and the isolated plasmid clones could encode PhoP-regulated genes that are being expressed due to high-gene dosage.

1.3. Evaluating Peptide Resistance

Once isolated, mutants displaying an altered susceptibility profile need to be further characterized to ascertain the contribution of the affected gene product towards resistance, information that is key to understanding the resistance mechanism. In this regard, both gradient plates and liquid assays are particularly useful. The gradient plate assay allows the side by side comparison of susceptibility among different strains. The liquid assay can also be used to determine the minimal inhibitory concentration (MIC), but is particularly useful for comparing the rate of loss of viability under defined and varied incubation conditions (*see* **Subheading 3.**).

The gradient plate assay has a limitation in that it does not distinguish loss of viability as resulting from a bacteriostatic or lethal event. A cell is considered viable if it can grow and form a visible colony (or streak) on a solid medium—which is the basis of the plate count method. Unfortunately, the absence of visible growth does not distinguish between dead cells and cells that are still metabolically alive but unable to divide and form colonies. Furthermore, this method cannot provide information as to whether the "lethal" event occurred immediately after protamine treatment or sometime during incubation on the agar medium. It is possible to determine the immediate effects of an antimicrobial peptide by using the fluorescence viability staining method that allows a real time distinction between live and dead cells. We have used this method to examine the bactericidal effects of protamine on hypersusceptible strains of *S. typhimurium*.

2. Materials

1. Defined medium for liquid assays: 50 m*M* MOPS, 4 m*M* tricine, 10 m*M* $(NH_4)SO_4$, 5 m*M* Na_2HPO_4, 1 m*M* KCl, 1 m*M* $MgSO_4$, 10 μ*M* $FeSO_4$, 0.2% glucose. A base solution is prepared by mixing the MOPS, tricine and $FeSO_4$, adjusting the pH to 7.0 with NaOH, and then filter sterilizing. All other components are stored as individual stock solutions and are added aseptically when fresh medium is needed. The $(NH_4)SO_4$, KCl, and $MgSO_4$ stock solutions are prepared at 1*M*; the Na_2HPO_4 at 0.5*M* with the pH adjusted to 7.0 with NaOH; and the glucose as a 20% solution.
2. Luria Bertani agar (for gradient plates): 1% tryptone, 0.5% yeast extract, 1% NaCl, 1.5% agar. The final pH is 6.8–7.2 and does not require adjustment. This medium can be autoclaved. For gradient plates, protamine sulfate is added to 2 mg/mL after autoclaving and cooling to 50°C. When stored at 4°C, the gradient and LB-agar plates are stable for 1 wk and 1 mo, respectively.
3. Protamine-LB agar for screening: Luria Bertani (LB) agar plus 1.2 mg/mL protamine sulfate. Autoclave the LB agar and cool to 50°C. With constant mixing, add protamine to a final concentration of 1.2 mg/mL. Pour approx 20

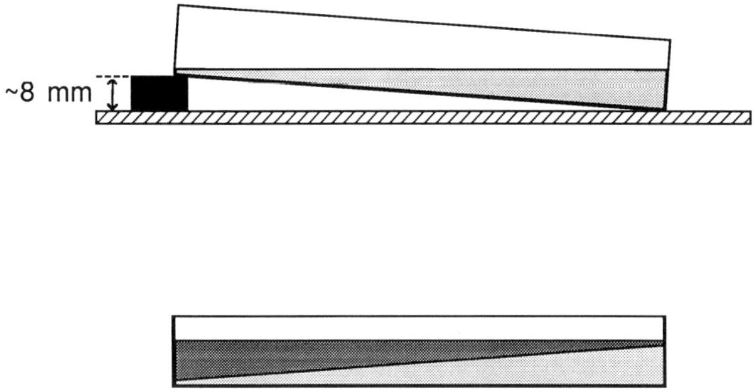

Fig. 1. Preparation of protamine gradient plates. Top: A bottom layer of plain LB agar is poured onto a square petri dish that is raised 8 mm on one side. Bottom: After the plain LB agar solidifies, the dish is placed on an even surface and a top layer of protamine LB agar is poured on top and let solidify.

mL each into 100 × 15 mm Petri plates. This medium is stable for at least 1 mo when stored at 4°C.

4. The protamine sulfate (5000–10,000 Dalton) can be purchased from Calbiochem (San Diego, CA). This compound appears to remain stable for up to a year when stored desiccated at 4°C. The stock solution used for the defined medium liquid assay is prepared fresh at a concentration of 2 mg/mL in sterile MOPS buffer base solution (*see* **step 1**, above). This solution can be kept at room temperature for several hours without loss of activity.
5. The fluorescence viability stain is sold as the LIVE/DEAD·*Bac*Light™ Viability Kit by Molecular Probes (Eugene, OR). This kit includes a mixture of two nucleic acid stains and should be stored in the dark at –20°C. These stains are toxic and latex gloves should be worn when handling it.

3. Methods

3.1. Preparation of Gradient Plates

1. Prepare two flasks of LB agar, 100 mL each, and autoclave. Each flask should contain a stir bar to facilitate mixing after autoclaving.
2. Cool to 50°C. With a sterile pipet transfer 25 mL each to four 100 × 100 mm square Petri plates. Swirl slightly to distribute the agar then elevate one side of the plate and let harden (*see* **Fig. 1A**). This forms the bottom layer of the gradient plate.
3. With constant mixing, add protamine to 2.0 mg/mL to the remaining 100 mL of agar. Continue to mix for 1–2 min.
4. Lay the Petri plates, with the bottom layer of LB agar, flat on the bench top. With a pipet, transfer 25 mL of protamine-LB-agar onto the bottom layer **(Fig. 1B)**. Swirl slightly to distribute and let harden.

3.2. Gradient Plate Assays

1. Inoculate 2-mL vol of LB broth (Luria Bertani without the agar) with the bacterial strains that are to be tested on the gradient plates. Incubate overnight at 37°C with shaking.
2. Dip a sterile swab in the overnight culture and roll out any excess liquid against the side of the test tube. Carefully streak the agar surface along the full length of the plate. Streak the full length 5–6 times to ensure even distribution of cells.
3. Incubate the plates in an inverted position at 37°C for 18–24 h.
4. The minimal inhibitory concentration (MIC) is calculated by dividing the length of the growth streak by the total length of the plate (90 mm) and multiplying by 2 mg/mL, the protamine concentration.

3.3. Determination of Viability by Fluorescence Staining

These assays are done in defined medium (pH 7.0) containing 11 m*M* glucose and 1 m*M* KCl. Incubation is at 37°C with rotary shaking at 200 rpm on a New Brunswick Scientific (Edison, NJ) Gyratory Water Bath Shaker. Refer to **Subheading 4.** for further details and a description of variations of this assay.

1. Inoculate 10 mL of fresh medium with 100 µL of a culture grown overnight in defined medium containing 11 m*M* glucose and 5 m*M* KCl. Incubate at 37°C with rotary shaking at 200 rpm and grow to OD_{600} 0.3.
2. At room temperature, dilute an aliquot of the growing culture into 5 mL of fresh medium, in a 50-mL Erlenmeyer flask, to a final OD_{600} of 0.1. For *S. typhimurium*, this results in a cell density of ca. 1×10^8 CFU/mL. Five milliliters are used initially to allow for the removal of 1-mL aliquots for determination of optical density.
3. Incubate at 37°C with rotary shaking at 200 rpm for 15–20 min. During this time, the OD_{600} should be monitored to ensure that the cells are growing. When the OD_{600} reaches 0.15, remove excess culture to achieve a final volume of 2 mL.
4. Transfer a 100-µL sample (the T_0 sample) to a 1.5-mL Eppendorf tube, add 1 µL of fluorescent stain (mixed according to the kit instructions), vortex briefly, and incubate in the dark at 37°C for 5–10 min.
5. To the remaining culture, add protamine to the desired concentration—which may range from 10–50 µg/mL depending on strain susceptibility.
6. After 20 min, remove another 100-µL sample and incubate with fluorescent stain as described above. Repeat this procedure for up to 3 h (or as long as desired).
7. To assess viability, transfer 7 µL of stained culture to a slide and cover with a glass cover slip. This step should be done quickly with as little exposure to light as possible. Observe under epifluorescence. The effect of incubation time and protamine concentration on viability is determined by counting the number of dead (red) and live (green) cells per field of view.

4. Notes

1. Transposons generally make polar mutations. Therefore, a mutant phenotype may be due to lack of expression of genes present downstream of the gene inactivated by the transposon. To establish that a particular gene is responsible for the phenotype of interest one needs to carry out complementation assays to demonstrate that a wild-type copy of the mutated gene is sufficient to restore wildtype levels of resistance to the mutant strain.
2. Once we establish that a transposon insertion is responsible for an altered peptide susceptibility phenotype, we are often anxious to identify the mutated gene. Therefore, we first clone the fragment harboring a segment of the transposon and the *Salmonella* chromosomal DNA at the site of transposon insertion (referred to as the joint fragment). We use this plasmid clone as a template in DNA sequencing reactions with a primer complementary to the end of the transposon. By searching the databases for genes or proteins related to the generated sequence (usually coming from a short run of 150–200 bases), we can establish whether the transposon mutated a known gene, a homolog of a gene that has already been described in another organism, or a novel sequence. This analysis can not only reveal the identity of the resistance determinant but also facilitates mapping and the cloning of the wild-type copy of the gene ***(17,18)***.
3. It is often important to establish whether two mutations map to different positions within the same locus and whether certain loci are tighlty linked. If the mutations were generated with the same transposon, it is necessary to convert this transposon into a derivative that confers resistance to a different antibiotic. For example, we have converted the MudJ transposon (which confers resistance to kanamycin) into Mud-Cam and Mud1-8 derivatives (which confer resistance to chloramphenicol and ampicillin, respectively). The resulting strains then allowed us to carry out phage transduction experiments to establish linkage between MudJ and Mud-Cam insertions in different peptide-resistant determinants.
4. The pH and ionic strength of the liquid media can be altered to enhance or reduce susceptibility to protamine. In general, cells are much more susceptible at pH 7.7 than at pH 6.3 and in media of low cation concentration (e.g., low K^+, Mg^{+2}).
5. The interaction of protamine with LB-agar ingredients would turn the agar an opaque, creamy/brown. This would make the bacterial growth streak very difficult to see. Visualization of growth could be enhanced by staining with a dilute India ink-distilled water solution. Add two drops India ink to 10 mL water and mix. Flood the plate with 5 mL of this solution and let stand for 5–10 s; then pour off. The result will be a slightly lighter growth streak with a distinct border against a darker agar background.
6. Protamine has broad spectrum antimicrobial activity and we found it was not necessary to filter sterilize the protamine stock solutions used in the liquid assays. Furthermore, the peptide could be added directly in powder form to the LB agar

medium used for preparation of the gradient plates. These plates showed no sign of contamination after nearly a month in cold storage or when left at room temperature for over 1 wk.

Acknowledgments

Resistance to antimicrobial peptides has been investigated thanks to financial support from the NSF, NIH, and the Monsanto/Searle-Washington University School of Medicine Biomedical Agreement to EAG. EAG is a recipient of a Junior Faculty Research Award from the American Cancer Society.

References

1. Zasloff, M. (1992) Antibiotic peptides as mediators of innate immunity. *Curr. Opin. Immunol.* **4,** 3–7.
2. Boman, H. G. (1995) Peptide antibiotics and their role in innate immunity. *Annu. Rev. Immunol.* **13,** 61–92.
3. Groisman, E. A. (1994) How bacteria resist killing by host-defense peptides. *Trends Microbiol.* **2,** 444–449.
4. Groisman, E. A., Parra, C. A., Salcedo, M., Lipps, C. J., and Heffron, F. (1992) Resistance to host antimicrobial peptides is necessary for *Salmonella* virulence. *Proc. Natl. Acad. Sci. USA* **89,** 11,939–11,943.
5. Christensen, B., Fink, J., Merrifield, R. B., and Mauzerall, D. (1988) Channel-forming properties of cecropins and related model compounds incorporated into planar lipid membranes. *Proc. Natl. Acad. Sci. USA* **85,** 5072–5076.
6. Kagan, B. L., Selsted, M. E., Ganz, T., and Lehrer, R. I. (1990) Antimicrobial defensin peptides form voltage-dependent ion-permeable channels in planar lipid bilayer membranes. *Proc. Natl. Acad. Sci. USA* **87,** 210–214.
7. Cruciani, R. A., Barker, J. L., Zasloff, M., Chen, H.-C., and Colamonici, O. (1991) Antibiotic magainins exert cytolytic activity against transformed cell lines trough channel formation. *Proc. Natl. Acad. Sci. USA* **88,** 3792–3796.
8. Cociancich, S., Ghazi, A., Hetru, C., Hoffmann, J. A., and Letellier, L. (1993) Insect defensin, an inducible antibacterial peptide forms votage-dependent channels in MIcrococcus luteus. *J. Biol. Chem.* **268,** 19,239–19,245.
9. Lehrer, R. I., Barton, A., Daher, K. A., Harwig, S. S., Ganz, T., and Selsted, M. E. (1989) Interaction of human defensins with *Escherichia coli*. Mechanism of bactericidal activity. *J. Clin. Invest.* **84,** 553–61.
10. García Véscovi, E., Soncini, F. C., and Groisman, E. A. (1996) Mg^{2+} as an extracellular signal: environmental regulation of Salmonella virulence. *Cell* **84,** 165–174.
11. Leyer, G. J., and Johnson, E. A. (1993) Acid adaptation induces cross-protection against environmental stresses in *Salmonella typhimurium. Appl. Environ. Microbiol.* **59,** 1842–1847.
12. Foster, J. W. and Spector, M. P. (1995) How *Salmonella* survive against the odds. *Annu. Rev. Microbiol.* **49,** 145–174.

13. Miller, J. H., ed. (1991) *Bacterial Genetic Systems. Methods in Enzymology.* Academic, San Diego.
14. Mäkelä, P. H., Sarvas, M., Calagno, S., and Lounatmaa, K. (1978) Isolation and characterization of polymyxin-resistant mutants of *Salmonella. FEMS Microbiol.* **3,** 323–326.
15. Roland, K. L., Martin, L. E., Esther, C. R., and Spitznagel, J. K. (1993) Spontaneous *pmrA* mutants of *Salmonella typhimurium* LT2 define a new -two-component regulatory system with a possible role in virulence. *J. Bacteriol.* **175,** 4154–4164.
16. Roland, K. L., Esther, C. R., and Spitznagel, J. K. (1994) Isolation and characterization of a gene, *pmrD,* from *Salmonella typhimurium* that confers resistance to polymyxin when expressed in multiple copies. *J. Bacteriol.* **176,** 3589–3597.
17. Parra-Lopez, C., Baer, M. T., and Groisman, E.A. (1993) Molecular genetic analysis of a locus required for resistance to antimicrobial peptides in *Salmonella typhimurium. EMBO J.* **12,** 4053–4062.
18. Parra-Lopez, C., Lin, R., Aspedon, A., and Groisman, E.A. (1994) A *Salmonella* protein that is required for resistance to antimicrobial peptides and transport of potassium. *EMBO J.* **13,** 3964–3972.

15

Assay of Antibacterial Activities of the Bactericidal/Permeability-Increasing Protein in Natural Biological Fluids

Jerrold Weiss

1. Introduction

All living organisms are constantly confronted by invading microbes. In mammals, host defense against microbial invasion requires specialized cellular and extracellular elements in blood and a complex signalling machinery that triggers their mobilization to extravascular sites of invasion in a highly regulated manner. This process is driven by microbe-specific signals, including highly conserved and chemically unique microbial (envelope) components, such as Gram-negative bacterial lipopolysaccharides (LPS) *(1,2)*.

The neutrophilic polymorphonuclear leukocyte (PMN) plays an essential role in defense against bacterial invaders. As originally anticipated, a search in these cells has revealed a broad spectrum of "endogenous antibiotics" *(3)*. Among the specific cytotoxic products of the PMN is the so-called bactericidal/permeability-increasing protein (BPI), a highly cationic 50- to 55-kDa granule-associated protein with highly selective and potent activity against Gram-negative bacteria (GNB) *(4)*. The fates of BPI-sensitive bacteria exposed to isolated BPI or ingested by intact PMN are similar *(5,6)*, suggesting an important role of BPI in the intracellular action of PMN against this group of microorganisms. The selective and potent toxicity of BPI toward GNB reflect high affinity interactions with the highly conserved lipid A (inner core) region of LPS that are unique to the GNB outer envelope *(1,4)*. These interactions occur both in the bacterial envelope and with cell-free dispersed LPS, leading not only to potent antibacterial cytotoxicity but also to inhibition of the endo-

From: *Methods in Molecular Biology, Vol. 78: Antibacterial Peptide Protocols*
Edited by: W. M. Shafer, Humana Press Inc., Totowa, NJ

toxic activity of LPS *(4,7,8)*. BPI may thus contribute both to elimination of bacteria that produce LPS and suppression of the proinflammatory activity of LPS already present. Both of these functions may contribute to control of LPS-driven inflammatory reactions that can have life-threatening consequences if left unabated *(1)*.

These dual properties of BPI have prompted examination of the endogenous mobilization of BPI in extracellular fluids during inflammation *(9–11)* and the exogenous application of recombinant BPI derivatives as a novel therapeutic approach to invasive GNB infections when endogenous defenses are limiting *(12)*. A meaningful evaluation of the ability of BPI (or BPI derivatives) to act as extracellular bactericidal and/or endotoxin-neutralizing agents has required the development of in vitro assays that more closely mimic the conditions of complex biological fluids. Historically, in vitro characterization of most antimicrobial PMN proteins and peptides has relied on experimental test systems that maximized assay sensitivity and convenience and focussed on the action of an individual agent in isolation *(3,6,13)*. By examining BPI action in both artificial and more complex biologically relevant media, we have identified several positive and negative modulators of BPI action in biological fluids, some present constitutively and others mobilized during inflammation (*see* **Fig. 1**). For example, initial binding of BPI to target bacteria is competitively inhibited by physiological extracellular concentrations of divalent cations (Mg^{2+} and Ca^{2+}; approx 1.5–3 m*M*) when BPI is tested at the n*M* concentrations that are achieved in local inflammatory fluids or during exogenous administration *(11,12)*. In contrast, under the same conditions, BPI binding and neutralization of purified LPS is apparently unaffected by Mg^{2+} or Ca^{2+} *(14)*. The major protein in extracellular fluids, plasma-derived albumin, does not interfere with binding of BPI to target bacteria (or to cell-free LPS) but prevents progression of BPI-mediated bacterial injury from the outer membrane to the inner (cytoplasmic) membrane *(15)*. As a result, BPI-treated bacteria are not killed but rather retain metabolic activity and ultimately repair sublethal outer membrane damage. Many biochemical and physiological assays have been described that delineate the sublethal and lethal effects of BPI on the bacterial outer and inner envelope, respectively *(15,16)*. However, the simplest test is to assay bacterial colony-forming ability in bacteriologic media (e.g., nutrient agar) with and without supplemented albumin (≥1 mg/mL;*15*). Inhibition of bacterial colony formation in media containing albumin corresponds to lethal effects, whereas inhibition observed only in media without albumin corresponds to sublethal effects (that presumably progress to lethal injury in the absence of albumin).

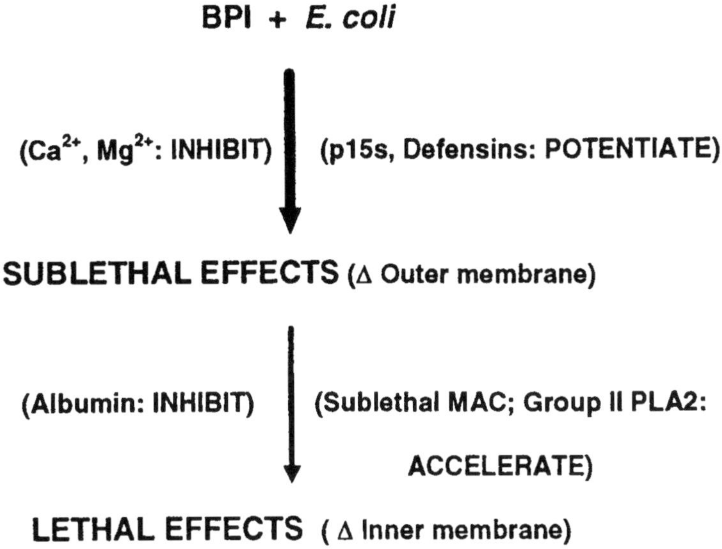

Fig.1. Steps in bactericidal action of BPI: Positive and negative modulators (*see text* for details).

1.1. Assay of BPI Activity in Whole Blood/Plasma

Despite the constitutive presence in biological fluids of agents that together can inhibit both the initiation of BPI antibacterial action (Ca^{2+}, Mg^{2+}) and its progression to lethal damage (albumin), addition/mobilization of n*M* concentrations of BPI in biological fluids confers potent bactericidal activity toward BPI-sensitive Gram-negative bacteria *(7,11)*. Detection of BPI-dependent antibacterial activity in either plasma or inflammatory fluids has been facilitated by using target organisms that express surface components (e.g., capsular polysaccharides, specific outer membrane proteins) that confer high levels of resistance to constitutively present extracellular bactericidal systems (e.g., complement) but do not increase bacterial resistance to BPI *(17,18)*. BPI-dependent bactericidal activity in plasma and inflammatory fluids is greater toward bacteria expressing LPS with short (R-chemotype) polysaccharide chains consistent with the inhibitory effect of long polysaccharide chains (O-antigen) on BPI interactions with intact bacteria *(7,18)*.

Identification of plasma component(s) that promote BPI action and overrride the inhibitory effects of divalent cations and albumin under physiologic conditions was initially complicated by the modifying effects of anticoagulants routinely used for preparation of plasma in vitro. Thus, the polyanion heparin competitively inhibits BPI binding to both intact bacteria and isolated LPS, whereas citrate and EDTA reduce the levels of divalent cations and, hence, artifactually promote BPI interaction with bacteria. Moreover, by reducing levels of Ca^{2+} (and Mg^{2+}), citrate and EDTA inhibit activation of the complement system. In the absence of anticoagulants, the fluid (serum) collected after sedimentation of cells, cellular debris, and fibrin clots contains substantially elevated levels of normally intracellular granule-associated proteins, including BPI, released from cells (e.g., PMN, platelets) during in vitro coagulation and cell activation. These problems can be circumvented by using an alternative anticoagulant, hirudin. Hirudin is a small protein product of leeches that promotes bleeding by inhibiting the catalytic action of thrombin ***(20)***. At concentrations required to prevent visible clotting of blood during incubations at 37°C for up to 5 h, hirudin affects neither the bactericidal activity of serum (complement) tested against complement-sensitive rough strains of *Escherichia coli* and *Salmonella typhimurium* nor that of BPI tested in buffered balanced salts solutions with and without Ca^{2+} and Mg^{2+}. In both plasma and whole blood containing hirudin, added BPI produces potent bactericidal effects against encapsulated *E.coli* that are resistant to the effects of plasma (e.g., complement) and whole; blood (e.g., phagocytes and complement) in the absence of BPI (**Fig. 2A**; *see* **Note 1**).

The use of whole blood also permits assay of the ability of BPI to inhibit the cytokine-inducing activity of the added bacteria (**Fig. 2B**; **ref.** *7*). This can be readily monitored by measuring the extracellular accumulation of tumor necrosis factor (TNF-α) during incubations of whole blood with bacteria (±BPI; **ref.** *7*; *see* **Notes 16–18**).

1.2. Assay of BPI Activity in Serum

The use of hirudin provides an experimental setting in which the antibacterial action of BPI in plasma and whole blood ex vivo can be assessed without introducing additives that demonstrably modify the activity of BPI and/or of potential modulators of BPI activity present in these biological fluids (suspensions). However, routine use of hirudin may be curtailed by the high cost and limited availability of the natural and recombinant products. Therefore, an alternative setting to assay BPI activity would be advantageous. As mentioned above, in comparison to normal plasma, serum contains higher levels of many normally granule-associated antimicrobial proteins, such as BPI ***(23,24)***. Most of these proteins and peptides are highly cationic in nature and can be selectively adsorbed to cationic exchange resins under physiological salt conditions. Recovered unbound proteins ("serum unbound") represent >99% of the total serum protein and contain the same bactericidal activity as the starting serum toward complement-sensitive rough strains of *E.coli* and *S. typhimurium*

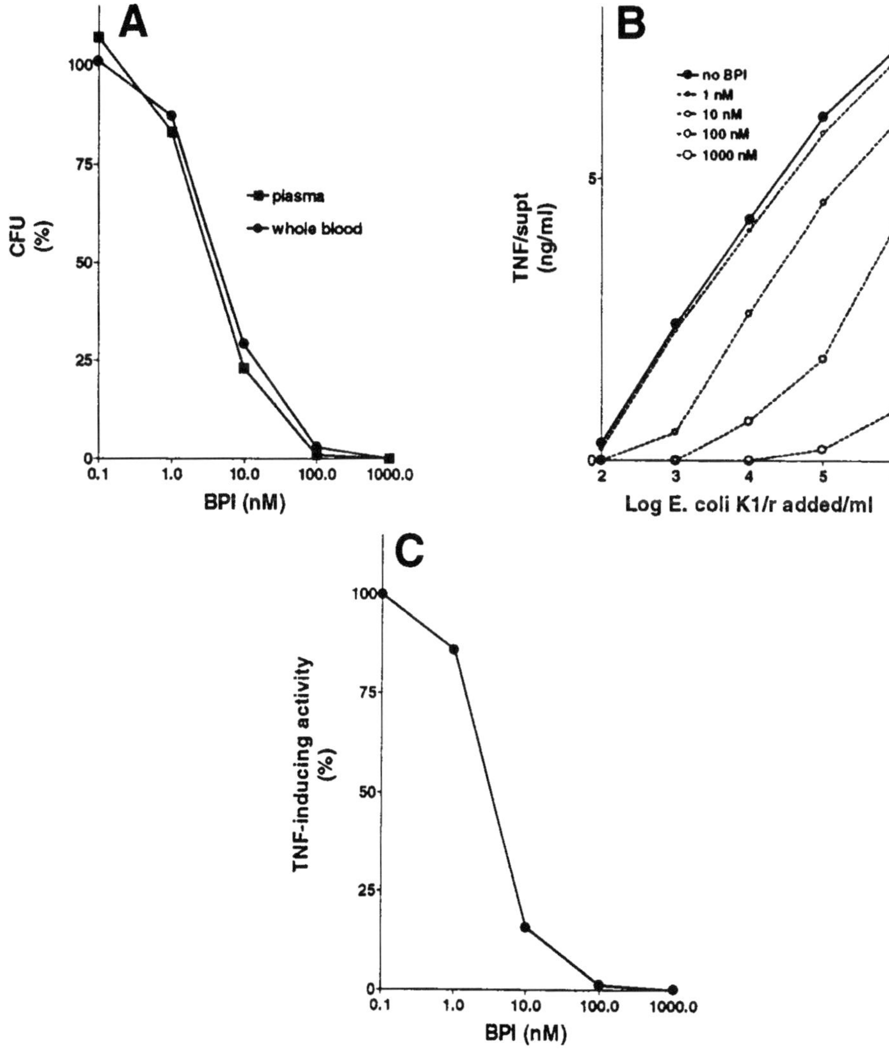

Fig. 2. Antibacterial activity of BPI in whole blood and plasma collected in hirudin. (**A**) Bactericidal activity of BPI toward *E. coli* K1/r (10^6/mL) during 60 min incubation. (**B**) TNF-inducing activity of *E. coli* K1/r during 5 h incubation in whole blood: Effect of BPI. (**C**) Transformation of data in panel B to represent ability of BPI to inhibit bacterial TNF-inducing activity. Note that 0.1 n*M* BPI had no effect on either bacterial viability or TNF-inducing activity.

(**Fig. 3A**). However, cationic proteins such as BPI are quantitatively adsorbed to the resin. The recovered unbound fraction exhibits reduced activity toward BPI-sensitive encapsulated *E.coli* (**Fig. 3B**) reflecting depletion of BPI. Addi-

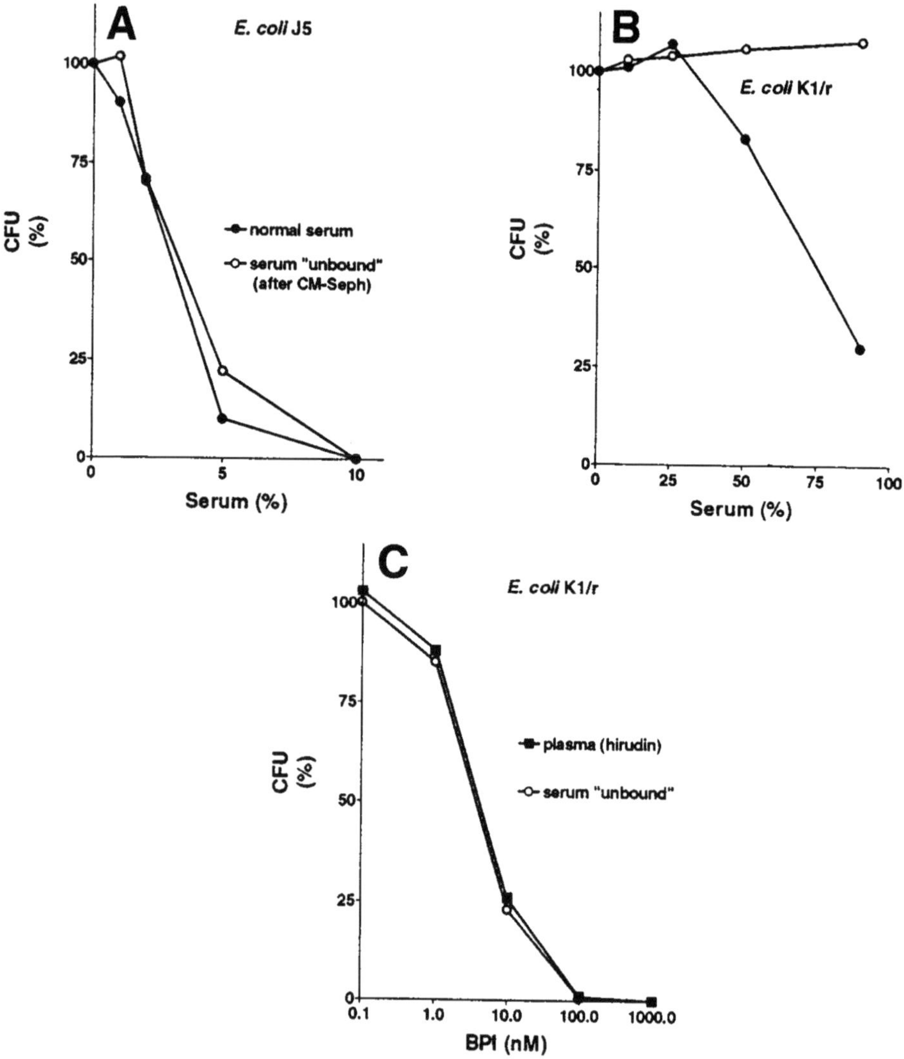

Fig. 3. Comparison of the bactericidal activity of serum toward *E. coli* J5 (**A**) and K1/r (**B**) before and after adsorption of serum to CM-Sephadex. Comparison of activity of human BPI toward *E. coli* K1/r in plasma collected in hirudin and in serum "unbound" (**C**). All incubations contained 10^6 bacteria/mL. *See text* for additional details.

tion of BPI to the "serum unbound" fraction confers potent bactericidal activity toward encapsulated *E.coli* that is identical to BPI activity in plasma prepared from whole blood containing hirudin (**Fig. 3C**; *see* **Notes 4** and **5**).

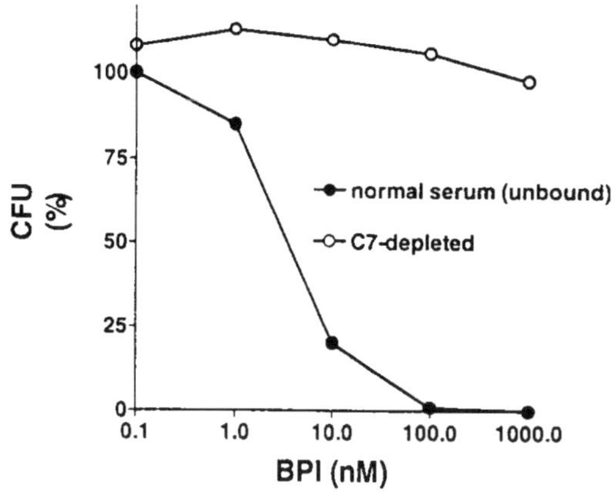

Fig. 4. Bactericidal activity of added BPI in normal serum (unbound) or C7-depleted serum. Incubations contained 10^6 bacteria/mL *E. coli* K1/r and were carried out for 60 min before assay of bacterial viability.

The use of "serum unbound" as a medium to mimic plasma for the study of BPI action has enabled us to make use of commercially available (e.g., Quidel, San Diego, CA) sera with defined deficiencies of specific complement proteins to determine the role of complement in BPI action in biologically fluids. As shown in **Fig. 4**, BPI-dependent bactericidal activity toward encapsulated *E. coli* is markedly reduced in the absence of C7 (or C6, 8 or 9; not shown). Thus the bactericidal action of BPI toward encapsulated *E. coli* in "serum unbound" (plasma) is dependent on its ability to act in synergy with nonlethal assemblies of the membrane attack complex of complement (*see* **Note 2**).

1.3. Mobilization and Action of BPI in Inflammatory Fluids

An important limitation of studies of extracellular BPI action in plasma or serum is that the composition of these fluids reflects what is present under "resting" conditions in the animal whereas, during inflammation, the composition and functional properties of these fluids change in a way that reflects the dynamics of the inflammatory response. Moreover, much of inflammation normally occurs at extravascular sites and thus the composition of these fluids is only in part derived from plasma via transudation. Under resting conditions, BPI levels in extracellular fluids are very low (≤5 ng/mL) ***(23)***. Hence, the potential extracellular action of BPI either during exogenous therapeutic administration or endogenous extracellular mobilization is likely to be limited to sites of inflammation. A particularly amenable experimental setting in which to study the mobilization and function of extracellular BPI during inflammation is by the experimental induction of sterile inflammatory peritoneal exudates in New Zealand White rabbits (*see* **Note 14**). This procedure has been used for decades to induce nearly homogeneous PMN-rich exudates ***(25)***. The inflammatory response that is induced shows many of the hallmarks of acute inflammation induced by injury or infection and with time (>18 h) also manifests characteristics of more chronic inflammation (e.g., increased accumulation of mononuclear leukocytes). Much of the machinery involved both in cytotoxic action against Gram-negative bacteria and recognition and response to LPS is closely similar in rabbits and humans making the rabbit a good experimental animal model to study host responses to Gram-negative bacteria. Use of this model, in fact, provided the setting in which BPI was first identified and its extracellular mobilization and function first demonstrated. Study of the inflammatory (ascitic) fluid has also revealed additional positive modulators of BPI action that are mobilized to high extracellular concentrations during inflammation. These include granule-derived proteins of PMN that potentiate the early sublethal action of BPI (e.g., p15s; **refs. *11*** and ***26***) and proteins that accelerate the transition from sublethal to lethal damage (e.g., Group II 14-kDa phospholipase A_2; **ref. *6***).

2. Materials

2.1. Whole Blood and Plasma

1. Normal human blood (≤10 mL).
2. Polypropylene tubes, sterile 5-mL round-bottomed tubes with snap caps.
3. Hirudin (Sigma, St. Louis, MO).
4. Eppendorf tubes, 0.5, 1.5 mL, sterile, polypropylene.
5. Transfer pipets, disposable, polyethylene (2 mL).

2.2. Serum

1. Normal human blood (≤10 mL).
2. Sterile tubes, no additive (Becton Dickinson Vacutainer system, Rutherford, NJ).
3. Pasteur pipets.
4. CM-Sephadex (Pharmacia, Piscataway, NJ).
5. 1.5*M* NaCl buffered with 2.5 m*M* Tris-HCl, pH 7.4.
6. Physiological saline (0.9%, w/v) buffered with 2.5 m*M* Tris-HCl, pH 7.4.
7. Complement (C6, 7, 8, or 9)-depleted human serum (Quidel, San Diego, CA).

2.3. Rabbit inflammatory (Ascitic) Fluid

1. New Zealand White rabbits (2–3 kg; ≥6 mo).
2. 60-mL Sterile syringes, monoject, sterile, nonpyrogenic (Becton-Dickinson).
3. 19-gage Needle, winged infusion set (Terumo Medical, Elkton, MD).
4. 16-gage Disposable hypodermic needles (Becton-Dickinson).
5. 0.9% Sodium chloride for irrigation (Baxter, Deerfield, IL).
6. Polypropylene, 50-mL sterile conical centrifuge tube.

2.4. Bacteria

1. Encapsulated *E. coli* (K1/r obtained from Alan Cross, University of Maryland at Baltimore School of Medicine, Baltimore, MD; O7K1 [ATCC 23505] obtained from American Type Culture Collection, Rockville, MD).
2. Trypticase soy broth (TSB), 30 g/L distilled water (Difco, Detroit, MI).
3. Nutrient agar (8 g nutrient broth, 5 g sodium chloride, and 13 g Difco Bitek agar per L of distilled water).
4. Bovine serum albumin.
5. Petri dishes (Falcon; 60 × 15 mm, polystyrene).
6. 25-mL Disposable sterile pipets.
7. Potato tubes, 50-mL, with caps, sterile.

2.5. TNF Assay

1. Biokine TNF test kit (T Cell Sciences, Cambridge, MA).
2. RPMI (Gibco-BRL, Gaithersburg, MD).

3. Methods

3.1. Collection of Whole Blood and Plasma in Hirudin

1. Collect blood from normal healthy volunteers into polypropylene tubes (5 mL) containing hirudin (final concentration, 20 U/mL).
2. Spin blood at 2000*g* for 10 min at room temperature to sediment blood cells.
3. Carefully remove supernatant (plasma); spin 10,000*g* for 20–30 min to remove residual debris (*see* **Note 3**).

3.2. Serum

3.2.1. Collection of Serum

1. Collect blood from normal healthy volunteers into Vacutainer tubes without additive.
2. Incubate blood for 30–60 min at room temperature to allow clot formation and sit in ice bath for an additional 30–60 min to promote clot retraction.
3. Pipet off upper soluble phase (serum).
4. Spin serum at 10,000*g* for 20 min to remove any debris (*see* **Note 3**).

3.2.2. Fractionation of Serum by CM-Sephadex Chromatography

1. Before collection of serum, prepare CM-Sephadex resin by swelling in 1.5*M* NaCl buffered with 2.5 m*M* Tris-HCl, pH 7.4, overnight at room temperature; wash the resin several times with 0.9% saline containing the same buffer to equilibrate the resin in the latter buffer.
2. Transfer the desired volume of resin in equilibration buffer to a conical polypropylene tube of appropriate volume and briefly spin (1000*g* for 30 s) to sediment the resin. Carefully remove excess buffer from resin.

3. Incubate serum with CM-Sephadex (0.1 mL resin/mL of serum) by rotation for 2 h at 4°C (e.g., cold room).
4. Sediment resin (as above) and recover supernatant (serum "unbound") (*see* **Note 4**).

3.3. Collection of Rabbit Inflammatory (Ascitic) Fluid (see Note 6)

1. Prepare a supersaturated solution of oyster glycogen (2.5 mg/mL) in sterile, pyrogen-free physiological saline.
2. Briefly incubate solution at 37°C.
3. Inject 250–300 mL of glycogen-saturated saline into the peritoneal cavity of a rabbit, using a 60-mL syringe that is simultaneously connected via a three-way stopcock to a bottle containing the injected solution and a winged infusion set to deliver intraperitoneally the glycogen-saline.
4. At various times after injection (*see* **Note 7**), the inflammatory peritoneal exudate is collected by insertion of a 16-gage needle into the peritoneal cavity. The rabbit is placed on a small wooden table with a hole placed in the middle to allow protrusion of the animal's abdomen. This permits access to the peritoneal cavity for insertion of the needle and collection of the exudate by gravity flow.
5. The inflammatory fluid is separated from cells in the exudate by sedimentation of cells at 100–200*g* for 5 min followed by centrifugation of the recovered supernatant at 20,000*g* for 10 min to remove particulate material (*see* **Notes 8–13**).

3.4. Bacteria

3.4.1. Growth of Bacteria

1. Bacteria are innoculated from frozen stock cultures (*see* **Subheading 4.**) into TSB and incubated overnight at 37°C.
2. Overnight bacterial cultures are diluted 1:100 in fresh TSB and incubated at 37°C for approx 4 h to late logarithmic to early stationary phase (*see* **Note 15**).
3. Bacteria are harvested by sedimentation at 10,000*g* for 4 min and resuspended in sterile physiological saline to the desired concentration (bacterial concentration at time of harvesting ~1–2 × 10^9/mL); bacterial concentration is determined by measuring the optical density of the bacterial suspension at 550 nm (0.1 OD = ~4 × 10^7 bacteria/mL).

3.4.2. Assay of Bacterial Viability

1. Typical incubation mixtures contain 10^4–10^7 bacteria/mL of HEPES- or phosphate-buffered biological fluid (e.g., plasma, serum, ascitic fluid) to mimic as closely as possible the conditions of the natural biological fluid.
2. After incubation for desired time, aliquots of the bacterial samples are serially diluted in sterile physiologic saline, mixed (10–50 µL) with approx 5 mL of molten (48°C) nutrient agar ± albumin (1 mg/mL) and added to a Petri dish (60 × 15 mm). The dishes are swirled to maximize mixing of the bacteria in the nutrient agar before solidification of the agar.

3. Bacterial colonies are enumerated after incubation of the nutrient agar at 37°C for 18–24 h.

3.5. Assay of TNF-Inducing Activity of Bacteria

1. Bacteria (10^2–10^7/mL) are added to whole blood (±BPI), 0.1 mL total volume, and incubated at 37°C.
2. After the desired incubation, samples are diluted four times with RPMI and spun at 1000*g* for 5 min to collect the extracellular medium.
3. TNF in the recovered extracellular medium is measured by ELISA following the instructions of the manufacterur of the test kit.

4. Notes

1. The bactericidal effects of added BPI are essentially the same in plasma and whole blood (*see* **Fig. 2A**), suggesting that added BPI principally promotes extracellular killing of these bacteria.
2. The activity of BPI is greatly reduced in preheated (56°C, 30 min) plasma consistent with an important role for complement in BPI-dependent killing of complement-resistant Gram-negative bacteria in plasma (and whole blood).
3. After collection, serum (and plasma) should be stored at –70° to preserve complement-dependent activity. Samples should be used after no more than 1 cycle of freezing and thawing.
4. The efficacy of depletion of cationic proteins from serum can be monitored by immunoassay of specific proteins (e.g., BPI, p15s; **ref. *11***) or by assay of Group II phospholipase A_2 activity using autoclaved *E. coli* ***(24)***. If necessary, adsorption of the first recovered unbound fraction can be repeated once without loss of complement activity.
5. Pretreatment of plasma collected in hirudin with CM-Sephadex does not reduce either plasma activity toward complement-sensitive bacteria or BPI activity (in plasma) toward encapsulated *E. coli*.
6. Before the initial collection, animals are injected 2–3 times at 1- to 2-wk intervals with approx 50–100 mL of glycogen/saline to "prime" the response. After initiating collection of exudates, animals are challenged at 2- to 3-wk intervals and yield exudates that are highly reproducible in cellular yield and cellular and extracellular protein content and functional activities.
7. The inflammatory exudate can be collected at various times after injection of glycogen/saline including at several times from a single exudate. PMN influx is significant by 2–3 h (approx 10^6/mL). Further accumulation to $1–3 \times 10^7$ cells/mL (>95% PMN) occurs by 8–12 h. After 16 h, recovery of PMN diminishes, whereas accumulation of mononuclear leukocytes increases.
8. BPI-mediated activity in ascitic fluid is stable at 4°C for at least 3 mo.
9. BPI-dependent antibacterial activity is most readily monitored by testing the activity of the collected ascitic fluid against a complement-resistant encapsulated strain of *E.coli* that contains short-chain LPS (e.g., *E. coli K*1/r) and, hence, is

highly sensitive to BPI. Activity is also manifest toward encapsulated bacteria containing long-chain LPS (e.g., *E. coli* O7K1). To mimic as closely as possible the conditons of the natural inflammatory fluid, assays are carried out in undiluted ascitic fluid (buffered with 10 m*M* phosphate or HEPES [pH 7.4] to maintain pH) with increasing numbers of bacteria (e.g., 10^4–10^8 bacteria/mL) incubated for various times. Sublethal and lethal effects are routinely monitored by measuring bacterial colony formation in nutrient agar (±0.1% albumin) ***(15)***. The BPI-dependence of the observed antibacterial activity is most definitively determined by measuring antibacterial activity in the presence and absence of neutralizing antibodies to BPI.

10. Complement-mediated, BPI-independent antibacterial activity of the inflammatory fluid can be assessed by adsorption of BPI to CM-Sephadex as described above for treatment of serum. Unbound proteins (>99% of total ascitic fluid protein) contain all complement activity of the inflammatory fluid. If antimicrobial activity is detected in the recovered unbound fraction, the role of the membrane attack complex of complement can be determined by comparing the activity of unbound fractions of inflammatory fluids collected from normal rabbits and from C6-deficient rabbits. The latter rabbits can be obtained from animal facilities at the National Institutes of Health. General characteristics of the inflammatory exudate in C6-deficient rabbits (e.g., cellular and extracellular protein influx and accumulation) are indistinguishable from that of normal rabbits. If antibacterial activity is reduced in fractions of inflammatory fluid from C6-deficient animals, the role of C6 can be confirmed by testing the effect of added C6. Purified human C6 is commercially available (e.g., Quidel Laboratories) and functionally interchangable with rabbit C6.
11. Recovery in active form of proteins adsorbed to CM-Sephadex that are required for BPI-dependent antibacterial activity (e.g., BPI) can be best achieved by washing the resin twice with 0.1 vol of 1.5*M* NaCl containing 20 m*M* sodium acetate/acetic acid buffer, pH 4.0 ("high salt eluate"). Recovery of BPI, p15s, and phospholipase A_2 in the high salt eluate is approx 50%. After dialysis vs 20 m*M* acetate buffer, pH 4.0 and concentration to 0.1 vol of ascitic fluid, recombination of the high salt eluate with the flow-through fraction containing unbound proteins yields about half the bactericidal activity of the original ascitic fluid.
12. The contribution of macromolecular components (e.g., proteins) of the unbound fraction to BPI-dependent antibacterial activity of the whole ascitic fluid can be tested by preparing a protein-poor ultrafiltrate of the unbound fraction (or whole ascitic fluid) and comparing the antibacterial activity of the high salt eluate diluted in either the unbound protein fraction or the protein-poor ultrafiltrate of this fraction. Albumin (1% w/v) can be added to restore the bulk protein content of the ascitic fluid and unbound protein fractions. In this way, the bioassay reflects the normal electrolyte ± bulk protein composition of the natural inflammatory fluid. If the activity of the high salt eluate is greater in the presence of the unbound protein fraction, the possibility of synergy between components of

the high salt eluate (e.g., BPI) and of complement (e.g., the membrane attack complex) in the unbound protein fraction can be readily tested by making use of protein fractions derived from normal vs C6-deficient rabbits and purified C6 as described in **Note 10**.

13. The possible contribution of constituents of the high salt eluate other than BPI to the BPI-dependent of the high salt eluate can be assessed by comparing the antibacterial activity of the high salt eluate to that of a corresponding amount of purified BPI *(16)*. BPI content of the high salt eluate can be readily determined by immunoassay *(11)*. Further purification of the high salt eluate can be achieved by reversed-phase HPLC using a C4 column (e.g., Vydac, Hesperia, CA) resolved with a gradient of acetonitrile in 0.1% (v/v) triflouroacetic acid *(24)*. Recovered protein fractions can be lyophilized and resuspended in 20 m*M* acetate buffer, pH 4.0.
14. Similar approaches should be feasible with other inflammatory fluids including "wound" and blister fluids and plasma from patients or animal models of sepsis.
15. Encapsulated *E. coli* are grown to early stationary phase to maximize complement resistance.
16. The TNF-inducing activity of Gram-negative bacteria such as *E. coli* is remarkably potent *(7,22)*. Induction of TNF production/secretion is maximal at approx 10^6 bacteria/mL and detectable at 10^2–10^3 bacteria/ml (*see* **Fig. 2B**). Therefore, these assays should be carried out at $\leq 10^6$ bacteria/mL of blood to permit a quantitative assessment of the dose-dependent induction of TNF by the added bacteria and of the inhibitory effects of BPI. As little as 50 µL of blood is sufficient for each sample but incubations should be carried out for at least 3–5 h to permit sufficient extracellular accumulation of TNF.
17. The dose-dependence of bacterial TNF-inducing activity is non-linear (*see* **Fig. 2B**). Hence, the inhibitory effects of BPI (or other agents) on bacterial signalling is best assessed by comparing the levels of TNF produced by BPI-treated bacteria to that of a bacterial standard curve (in the absence of BPI). The data can then be transformed to represent the dose-dependent inhibitory effects of BPI on bacterial signalling (*see* **Figs. 2B,C**).
18. Accumulation of TNF in the extracellular medium can also be measured by assay of cytotoxicity of the medium toward TNF-sensitive cell lines (e.g., L929 fibroblasts; **ref.** ***21***).

References

1. Rietschel, E. T. and Brade, H. (1992) Bacterial endotoxins. *Sci. Amer.* **267,** 54–61.
2. Raetz, C. R. H., Ulevitch, R. J., Wright, S. D., Sibley, C. H., Ding, A., and Nathan, C. F. (1991) Gram-negative endotoxin: an extraordinary lipid with profound effects on eukaryotic signal transduction. *FASEB J.* **5,** 2652–2660.
3. Levy, O. (1996) Antibiotics proteins of polymorphonuclear leukocytes. *Eur. J. Haematol.* **56,** 263–277.

4. Elsbach, P. and Weiss, J. (1993) Bactericidal/permeability-increasing protein and host defense against Gram-negative bacteria and endotoxin. *Curr. Opin. Immunol.* **5,** 103–107.
5. Elsbach, P. and Weiss, J. (1992) Oxygen-independent antimicrobial systems of phagocytes, in *Inflammation: Basic Principles and Clinical Correlates.* Gallin, J. I., Goldstein, I. M., and Snyderman, R., eds.) Raven, NY, pp. 603–636.
6. Elsbach, P., Weiss, J., and Levy, O. (1994) Integration of antimicrobial host defenses: role of the bactericidal/permeability-increasing protein. *Trends in Microbiol.* **2,** 324–328.
7. Weiss, J., Elsbach, P., Shu, C., Castillo, J., Grinna, L., Horwitz, A., and Theofan, G. (1992) Human bactericidal/permeability-increasing protein and a recombinant NH_2-terminal fragment cause killing of serum-resistant Gram-negative bacteria in whole blood and inhibit tumor necrosis factor release induced by the bacteria. *J. Clin. Invest.* **90,** 1122–1130.
8. Marra, M. N., Wilde, C. G., Collins M. S., Snable J. L., Thornton M. B., and Scott R.W. (1992) The role of bactericidal/permeability-increasing protein as a natural inhibitor of bacterial endotoxin. *J. Immunol.* **148,** 532–537.
9. Calvano, S. E., Thompson, W. A., Marra, M. N., Coyle, S. M., Riesthal, H. F., Trousdale, R. K., Barie, P. S., Scott R. W., Moldawer, L. L., and Lowry, S. F. (1994) Changes in polymorphonuclear leukocyte surface and plasma bactericidal/permeability-increasing protein and plasma lipopolysaccharide binding protein during endotoxemia or sepsis. *Arch. Surg.* **129,** 220–226.
10. Opal, S. M., Palardy, J. E., Marra, M. N., Fisher, C. J. Jr., McKelligon, B. M., and Scott, R. W. (1994) Relative concentrations of endotoxin-binding proteins in body fluids during infection. *Lancet* **344,** 429–431.
11. Weinrauch, Y., Foreman, A., Shu, C., Zarember, K., Levy, O., Elsbach, P., and Weiss, J. (1995) Extracellular accumulation of potently microbicidal bactericidal/permeability-increasing protein and p15s in an evolving sterile rabbit peritoneal inflammatory exudate. *J. Clin. Invest.* **95,** 1916–1924.
12. Elsbach, P. and Weiss, J. (1995) Prospects for use of recombinant BPI in the treatment of gram-negative bacterial infections. *Inf. Agents & Actions* **4,** 102–109.
13. Elsbach, P. (1994) Bactericidal/permeability-increasing protein in host defence against gram-negative bacteria and endotoxin. *Ciba Foundation Symp.* **186,** 176–187.
14. Katz, S. S., Chen, K., Chen, S., Dorfler, M. E., Elsbach, P., and Weiss, J. (1996) Potent CD14-mediated signalling of human leukocytes by *Escherichia coli* can be mediated by interaction of whole bacteria and host cells without extensive prior release of endotoxin. *Infect. Immun.* **64,** 3592–3600.
15. Mannion, B. A., Weiss, J., and Elsbach, P. (1990) Separation of sublethal and lethal effects of the bactericidal/permeability-increasing protein on *Escherichia coli. J. Clin. Invest.* **85,** 853–860.
16. Weiss, J. (1994) Purification and assay of bactericidal/permeability-increasing protein. *Meth. Enzymol.* **236,** 173–196.

17. Weiss, J., Victor, M., Cross, A. S., and Elsbach, P. (1982) Sensitivity of K1-encapsulated *Escherichia coli* to killing by the bactericidal/permeability-increasing protein of rabbit and human neutrophils. *Infect. Immun.* **38,** 1149–1153.
18. Russo, T. A., Sharma, G., Weiss, J., and Brown, C. (1995) The construction and characterization of colanic acid deficient mutants in an extraintestinal isolate of *Escherichia coli* (O4/K54/H5). *Microb. Path.* **18,** 269–278.
19. Weiss, J., Beckerdite-Quagliata, S., and Elsbach, P. (1980) Resistance of gram-negative bacteria to purified bactericidal leukocyte proteins. Relation to binding and bacterial lipopolysaccharide structure. *J. Clin. Invest.* **65,** 619–628.
20. Dang, Q. D. and DiCera, E. (1994) A simple activity assay for thrombin and hirudin. *J. Prot. Chem.* **13,** 367–373.
21. Baarsch, M. J., Wannemuehler, M. J., Molitor, T. W., and Murtaugh, M. P. (1991) Detection of tumor necrosis factor alpha from porcine alveolar macrophages using an L929 fibroblast bioassay. *J. Immunol. Meth.* **140,** 15–22.
22. Levy, O., Ooi, C. E., Elsbach, P., Doerfler, M. E., Lehrer, R. I., and Weiss, J. (1995) Antibacterial proteins of granulocytes differ in interaction with endotoxin: comparison of bactericidal/permeability-increasing protein, p15s, and defensins. *J. Immunol.* **154,** 5403–5410.
23. White, M. L., Ma, J. K., Birr, C. A., Trown, P. A., and Carroll, S. F. (1994) Measurement of bactericidal/permeability-increasing protein in human body fluids by sandwich ELISA. *J. Immunol. Meth.* **167,** 227–235.
24. Weinrauch, Y., Elsbach, P., Madsen, L. M., Foreman, A., and Weiss, J. (1996) The potent anti-*Staphylococcus aureus* activity of a sterile rabbit inflammatory fluid is due to a 14-kD phospholipase A2. *J. Clin. Invest.* **97,** 250–257.
25. Hirsch, J. G. (1956) Phagocytin: a bactericidal substance from polymorphonuclear leukocytes. *J. Exp. Med.* **103,** 589–611.
26. Ooi, C. E., Weiss, J., Levy, O., and Elsbach, P. (1990) Isolation of two isoforms of a novel 15 kDa protein from rabbit polymorphonuclear leukocytes that modulate the antibacterial actions of other leukocyte proteins. *J. Biol. Chem.* **265,** 15,956–15,962.

16

Assay Systems for Measurement of Chemotactic Activity

H. Anne Pereira

1. Introduction

The traditional view of the polymorphonuclear leukocyte (PMN) or neutrophil has been a cell whose primary function was to ingest and kill bacteria. It is becoming increasingly clear that a number of these granule constituents that were originally studied because of their antibiotic activity may in fact have other important roles in various aspects of inflammation *(1)*. Accumulating evidence would support the concept that the PMN is a pivotal effecter cell in the immune response, capable of generating cytokines, chemokines, and a variety of mediators that orchestrate the process of inflammation *(2)*. Of the antibacterial proteins present within the human PMN, cationic antimicrobial protein of M_r 37 kDa (CAP37) and the defensins have been shown to have chemotactic activity for monocytes *(3,4)*. CAP37 is chemotactic at concentrations of 1.3×10^{-8} to ^{-10}M for human monocytes. It appears to have no effect on neutrophil migration. In addition to confirming the chemotactic activity of CAP37 for monocytes, Flodgaard et al. *(5)* have demonstrated that CAP37 has chemotactic activity for fibroblasts and effects maturation of monocytes into macrophages. Henson et al. *(6)* have shown that instillation of CAP37 into rabbit airways leads to a significant and selective migration of rabbit blood monocytes into the lung. The minimal neutrophil and lymphocyte migration into the lung that occurred in response to CAP37 was indistinguishable from that induced by the instillation of vehicle alone.

CAP37 is a single-chain glycoprotein consisting of 222 amino acids. Sequence analysis indicated that CAP37 bears very substantial similarities with serine proteases important in various aspects of inflammation. Highest homol-

From: *Methods in Molecular Biology, Vol. 78: Antibacterial Peptide Protocols*
Edited by: W. M. Shafer, Humana Press Inc., Totowa, NJ

ogy was with neutrophil elastase (45%) and proteinase 3 (42%) and to a lesser extent with cathepsin G. However, owing to replacement of two of the three highly conserved residues of the catalytic triad present in serine esterases, CAP37 lacks enzymatic activity *(1)*. Two other serine proteases, bovine thrombin *(7)* and a trypsin-like protease in guinea pig plasma *(8)*, have been shown to be chemoattractants. The nonenzymatic, serine esterase, scatter factor or hepatocyte growth factor has potent activity for epithelial cells and vascular endothelial cells *(9)*. Proteins such as α-casein, β-casein, and bee venom mellitin, which are cationic and contain hydrophobic sections, show chemotactic activity *(10)*.

Although it had been known from the time of Metchnikoff and Leber that leukocytes could migrate along a gradient in a directional manner, in vitro analysis of chemotaxis had been slow due to the lack of a suitable technique. All of this changed when, in 1962, Boyden developed his now famous in vitro assay for the measurement of chemotactic activity of substances in solution. Over the years, minimal modifications have been made with regard to the general design of the apparatus; however, the basic principle for measuring chemotaxis has remained the same. The apparatus consists of two chambers separated by a filter that contains pores of a known and uniform size. The upper (or top) chamber contains the cells, which are allowed to settle onto the upper surface of the membrane. The lower chamber contains the chemoattractant, whose activity is to be measured. The pore size of the membrane is chosen such that the holes are large enough for the cells to actively crawl through them but not so large that the cells can physically fall through into the lower chamber. The chambers are then incubated in a humidified atmosphere for a certain length of time, which depends on the type of cell used for the study. The neutrophil requires a shorter incubation time than do monocytes. Lymphocytes, endothelial cells, and smooth muscle cells require substantially longer incubation periods. At the end of the incubation time, the filters are removed, stained, and an assessment made of the chemotactic activity of the substance by one of two methods. The "leading front technique" measures the distance migrated by the cells into the filter in micrometers. The second method depends on counting the numbers of cells migrating all the way through to the under side of the filter. This gives a numerical estimate of the chemotactic activity of the chemoattractant in the lower chamber.

The more commonly used methods for measuring chemotaxis include the modified Boyden chamber assay, the multiwell chemotaxis chamber assay, Costar Transwell™ cell culture chambers, and the "under-agarose" method, among others. We have had consistently reproducible and accurate results with the single well or modified Boyden chamber using the leading front technique, and this has been the apparatus and method of choice for measuring chemotaxis in my laboratory. We have found it adaptable to the various cells and chemoattractants under study. However, the mutliwell chemotaxis chamber *(11)* and Transwell™ cell culture chambers and "under agarose" methods are preferred by many others and have distinct advantages in certain experimental situations.

An important consideration in performing in vitro chemotaxis assays is the choice of filter. When using the modified Boyden chamber, one has the choice of nitrocellulose filters or polycarbonate filters. Nitrocellulose filters are required if one selects the leading front technique. These filters are approx 150 µm thick and therefore the distance migrated into the filter can be estimated. The polycarbonate filters are much thinner (approx 10 µm) and are unsuitable for this method. However, polycarbonate filters can be used in assays in which the numbers of cells migrating through to the undersurface of the filter are counted. These filters are available in a wide range of pore sizes. The most appropriate filter for measuring neutrophil chemotaxis is one with a 3.0-µm pore size. When measuring monocyte or lymphocyte chemotaxis, the appropriate pore size is 5.0–8.0 µm. In addition, the Costar Transwell™ cell culture inserts with polycarbonate filters of 3.0, 5.0, and 8.0 µm have proved to be an extremely convenient method for measuring chemotaxis. They are particularly useful for long-term cultures in which one requires a sterile apparatus. This is not possible with either of the modified or multiwell chemotaxis chambers. One drawback with these disposable, sterile units remains their cost.

A very important consideration when measuring chemotaxis is the ability to differentiate between chemotaxis (i.e., directed movement along a concentration gradient) and chemokinesis (i.e., nondirectional movement in the absence of a chemical gradient). This is most conveniently determined using the "checkerboard assay" designed by Zigmond and Hirsch *(12)*. In the normal chemotactic assay, one adds the chemoattractant to the lower chamber only. In the checkerboard assay, a series of chambers are set up such that there is an increasing amount of chemoattractant in the lower chamber (as per usual); however, there is also added an increasing amount of chemoattractant to the upper chamber, which contains the cells. In this manner a series of chambers are set up where there gradually emerges a point at which there is no concentration gradient between upper and lower chambers. Since chemotaxis is defined as a directed motion along a concentration gradient, one is able to differentiate between true chemotaxis and purely random or nondirectional movement. A theoretical example is depicted in **Table 1**, in which boxes with horizontal shading represent chemotactic activity, and boxes with diagonal shading indicate chemokinesis.

Table 1
Determination of Chemokinesis by Checkerboard Assay (Distance Migrated in μm)

Concentration of chemoattractant above the filter	Concentration of chemoattractant below the filter				
	0	**1X**	**2X**	**3X**	**4X**
0	**25**	**43**	**85**	**120**	**105**
1X	26	**30**	48	79	90
2X	28	29	**68**	85	91
3X	27	30	39	**53**	93
4X	25	32	35	32	**45**

When we first attempt to identify a chemoattractant we perform an initial screen using the polarization assay ***(13)***. Leukocytes undergo a shape change from spherical to a triangular or polarized form when exposed to chemoattractants. This change in shape occurs within the first few minutes, and the cells can be fixed so as to maintain this altered asymmetric morphology. The numbers of cells exhibiting polarization then gives an index of the level of chemotactic activity. It is important to note, however, that this method does not distinguish between chemotaxis and chemokinesis; therefore, confirmation needs to be made using the modified Boyden chemotaxis chamber assay.

A final, but extremely important aspect in performing chemotaxis assays is to ensure that all reagents, buffers, and chemoattractants to be analyzed are endotoxin free. We use a commercial chromogenic limulus amebocyte lysate assay for this purpose.

2. Materials

2.1. Leukocyte Separation

1. Normal human blood sample (50–60 mL).
2. Sterile sodium EDTA tubes (Becton Dickinson, Rutherford NJ).
3. Lymphocyte separation medium (LSM, Organon Teknika, Durham, NC).
4. 0.01*M* Phosphate-buffered saline (PBS) containing 0.15*M* NaCl, pH 7.4 (8.5 g NaCl, 1.07 g Na_2HPO_4, 0.34g NaH_2PO_4 in 1 L endotoxin-free water, or purchase calcium and magnesium-free PBS, Cellgro-Mediatech, Herndon, VA.
5. Gey's balanced salt solution (Gibco, Grand Island, NY).
6. Endotoxin-poor (free) bovine serum albumin (Pierce, Rockford, IL).
7. 2% endotoxin-free BSA in Gey's buffer. Prepare fresh, just prior to experiment.
8. Polypropylene, 50-mL sterile conical centrifuge tubes.

9. Polypropylene, 15-mL sterile conical centrifuge tubes.
10. 25-, 10-, and 5-mL disposable sterile pipets.
11. Sterile Pasteur pipets.
12. Coulter counter or hemacytometer.

2.1.1. Monocyte Separation

1. All materials listed in **Subheading 2.1.**
2. Nycoprep specific gravity 1.068 (Nycomed Pharma, Oslo, Norway).

2.1.2. Lymphocyte Separation

1. All materials listed in **Subheading 2.1.**
2. T-cell enrichment column (R & D Systems, Minneapolis, MN).

2.1.3. Neutrophil Separation

1. All materials listed in **Subheading 2.1.**
2. Dextran T500 (3% in normal sterile saline, Pharmacia Fine Chemicals, Piscataway, NJ). Store refrigerated for up to 4 wk.
3. Normal saline (0.9% NaCl in sterile, endotoxin-free water or 0.9% sodium chloride for Irrigation, USP Baxter, Deerfield, IL).
4. Sterile endotoxin-free water.
5. 3.5% NaCl.

2.2. Endotoxin Detection

1. Limulus amebocyte lysate quantitative chromogenic assay (Biowhittaker, Walkersville, MD).
2. Sterile polystyrene test tubes (12 × 75 mm, Fisher Scientific).
3. Sterile 96-well flat-bottomed polystyrene microtiter plate with lid (Corning, Cambridge, MA).
4. Microtiter plate reader.
5. Sterile pipets.
6. Sterile, endotoxin-free water.
7. 25% acetic acid solution in sterile endotoxin-free water.
8. Reagents, chemoattractants, and buffers that will be used in all aspects of chemotaxis assays.

2.3. Polarization Assay

1. Leukocytes adjusted to between 3–5 × 10^6 cells/mL.
2. 8% w/v paraformaldehyde in PBS (4°C). Can be stored refrigerated up to 1 mo.
3. Gey's balanced salt solution (Gibco).
4. Sterile polystyrene test tubes (12 × 75 mm, Fisher).
5. Microscope slides and cover slips.
6. Phase-contrast microscope.

7. Chemoattractant for testing at 10X final concentration.
8. Positive control for neutrophils and monocytes (fMet-Leu-Phe $10^{-7}M$ for final concentration of $10^{-8}M$, Sigma) and IL-8 for lymphocytes (final concentration 10 ng/mL, R & D Systems, Minneapolis, MN).
9. PBS (4°C).

2.4. Chemotaxis Assays

2.4.1. Modified Boyden Chemotaxis Chamber Assay

1. Neuroprobe™ modified Boyden chemotaxis chambers (Costar-Nucleopore, Cambridge MA).
2. Cellulose nitrate filters, 13-mm diameter (Sartorius, Edgewood, NY, 8-µm pore size for monocytes and lymphocytes and 3-µm pore size for neutrophils; or Millipore, Bedford, MA, 8-µm pore size filters for monocytes and lymphocytes; and 3-µm pore size filters for neutrophils).
3. Pair of fine curved forceps.
4. Sterile Pasteur pipets.
5. Humidified incubator (5% carbon dioxide).
6. Microscope slides and cover slips.
7. Microscope.
8. Xylene.
9. Ethanol: Absolute, 95, 70, and 50%.
10. 1% Sodium acetate.
11. Auto Hematoxylin (Research Genetics, Huntsville, AL).
12. Permount (Fisher).
13. Staining chamber.
14. Leukocytes suspended at 2×10^6 cells/mL in 2% endotoxin-free BSA in Gey's buffer.
15. Chemoattractant to be tested at appropriate dilutions (10^{-6}–$10^{-10}M$) in 2% endotoxin-free BSA in Gey's buffer.
16. Appropriate positive controls (fMLP at $10^{-8}M$ for monocytes and neutrophils; IL–8 at 10 ng/mL for lymphocytes) in 2% endotoxin-free BSA in Gey's buffer.

2.4.2. Multiwell Chemotaxis Chambers

1. 48 Multiwell chemotaxis chamber (Neuro Probe, Cabin John, MD).
2. Cellulose nitrate filters (Sartorius 25 × 80 mm, 3- and 8-µm pore size to be used with neutrophils and monocytes, respectively).
3. Fine forceps (two pairs).
4. Q-tips.
5. HumidifiedIncubator (5% Carbon dioxide).
6. Microscope slides (50 × 75 mm).
7. Cover slips.
8. Microscope.
9. Phosphate-buffered saline.

10. Methanol.
11. Diff Quik Stain (Baxter Scientific Products, Grand Prairie, TX) or staining materials as listed in **Subheading 2.4.1.**
12. Chemoattractants to be tested at appropriate dilutions including an appropriate positive control as outlined for modified Boyden chemotaxis chamber assay, **Subheading 2.4.1., step 16**.
13. Leukocyte cell suspension as for modified Boyden chemotaxis chamber assay.

2.5 Chemokinesis: "Checkerboard Assay"

1. Boyden or multiwell chemotaxis chambers.
2. All items listed in **Subheading 2.4.1.** or **2.4.2.** depending on whether the Boyden or multiwell assay is used.

3. Methods

3.1. Leukocyte Separation

1. Collect blood from normal healthy volunteers into tubes containing EDTA (*see* **Note 1**).
2. Carefully, layer approx 15 mL of anticoagulated blood over 8 mL of LSM in a sterile 50 mL polypropylene centrifuge tube.
3. Centrifuge tubes at 1500*g* for 20 min at room temperature.
4. As soon as the centrifuge comes to a stop, remove tubes with care so as not to upset the gradient (*see* **Notes 2–4**).
5. Using this technique, the blood should separate such that the top layer (straw colored) will be plasma. Just below the plasma is a narrow band (white) consisting of mononuclear cells, i.e., monocytes and lymphocytes. Below this band is a mixture of LSM and platelets, and at the bottom of the tube is a fairly large pellet (red) that consists of red blood cells and neutrophils.
6. Depending on which type of leukocyte one requires proceed accordingly.

3.1.1. Monocyte Separation

1. Using a Pasteur pipet, remove approx 5mL of the mononuclear layer. This is most easily done by gently moving the Pasteur pipet in a circular motion around the circumference of this layer. Be careful not to go too deep into the LSM/platelet layer, since the more platelet contamination one gets, the more likely one is to get clumping of cells.
2. Transfer these cells into a 50-mL centrifuge tube and add approx 40–45 mL of PBS.
3. Centrifuge tube at 150*g* for 10 min at room temperature.
4. Aspirate and discard supernatant and gently resuspend pellet in 5 mL PBS. Cells can be resuspended by very gently pipetting "up and down" with a Pasteur pipet, or by vortexing very gently on the lowest setting. Make sure that there are no cell clumps.
5. Place 3 mL of Nycoprep 1.068 in a 15-mL centrifuge tube.

6. Layer the 5 mL of cell suspension from step 4 over the Nycoprep and centrifuge at 600*g* for 15 min at room temperature.
7. Monocytes will layer at the top, whereas lymphocytes will pellet at the bottom of the tube.
8. Remove monocyte layer with Pasteur pipette and wash once with PBS at 150*g* for 10 min.
9. Count cells on a Coulter counter or in a hemocytometer and adjust cells to a concentration of 2×10^6 cells/mL in Gey's buffer containing 2% endotoxin-free BSA (*see* **Note 5**).
10. Monocytes prepared in this manner are approx 90% pure.

3.1.2. Lymphocyte Separation

1. Prepare T-cell enrichment column according to the manufacturer's instructions, by washing column four times with 2 mL of 1X wash buffer, made by diluting 10X buffer provided with kit.
2. Aspirate the mononuclear layer as described above for monocyte isolation and proceed up to **step 3**.
3. Resuspend cells in 2 mL of the 1X wash buffer from the kit.
4. Load cells on to the column, and allow cells to remain on column for 10 min before beginning the elution step.
5. Elute cells using 2 mL of 1X wash buffer.
6. Centrifuge cells at 250*g* for 5–7 min and wash once with PBS.
7. Count cells on a Coulter counter or in a hemacytometer and resuspend cells in Gey's buffer containing 2% BSA to a final concentration of 2×10^6 cells/mL.
8. Lymphocytes prepared in this manner are 90–95% pure.

3.1.3. Neutrophil Separation

1. Remove plasma, mononuclear and LSM layers such that one is left with red blood cell-neutrophil pellet at bottom of centrifuge tube.
2. To the pellet of cells add approx 6 mL of dextran solution and 15–20 mL of PBS.
3. Gently invert and swirl tube so as to mix cells. Do not vortex.
4. Allow red blood cells to settle under gravity for 30–45 min at room temperature.
5. Red blood cells will rouleaux and settle to the bottom of the tube, leaving a light pink supernatant that consists mainly of granulocytes and a few red blood cells.
6. Remove the supernatant to a 50-mL centrifuge tube, and centrifuge at 150*g* for 10 min at room temperature.
7. In the meantime, add 12 mL endotoxin-free water to one centrifuge tube and 4 mL 3.5% NaCl to another tube and place these in an ice bucket.
8. To pellet from **step 6**, add the 12 mL ice-cold water and vortex gently for 10 s to lyse red blood cells.
9. Add the 4 mL ice-cold 3.5% NaCl to return solution back to isotonicity.
10. Centrifuge at 150*g* for 5 min at 4°C, and discard supernatant.
11. Gently resuspend cells (neutrophils) in PBS and wash twice more by centrifugation.

12. If red blood cell contamination is still high it may be necessary to perform a second lysis **(steps 9–11)**.
13. Count cells on coulter or on a hemacytometer and adjust cells to a concentration of 1×10^6 cells/mL in Gey's buffer containing 2% endotoxin-free BSA.

3.2. Endotoxin Detection

1. Read the instruction booklet provided by Biowhittaker for the QCL-1000 limulus amebocyte lysate kit. This provides excellent guidelines for performing the assay (*see* **Notes 6–8**).
2. Prepare endotoxin standards in sterile test tubes, using stock endotoxin provided in kit and sterile endotoxin-free water either provided with kit or purchased independently.
3. Set up the assay in the sterile microtiter plate as follows.
4. 50 µL each of test samples, blank and standards are added to the plate.
5. Place microtiter plate in 37°C water bath to allow all reagents to reach constant temperature.
6. Add 50 µL of limulus amebocyte lysate (LAL) reagent provided with kit.
7. Mix and incubate at 37°C for 10 min in water bath.
8. Add 100 µL of substrate solution provided with kit and incubate at 37°C for 6 min.
9. Add 100 µL of 25% acetic acid to stop the reaction. Mix well.
10. Make sure all moisture is removed from outside of plate.
11. Read the absorbance at 405 nm on a microtiter plate reader.
12. Calculate Endotoxin U/mL of test samples by reading values from standard curve.

3.3. Polarization assay

1. Adjust leukocytes to between $3–5 \times 10^6$ cells/mL in Gey's buffer (*see* **Note 9**).
2. Keep cells at room temperature for 15–30 min before the start of the assay.
3. Prepare test chemoattractant in Gey's buffer such that it is 10X final concentration.
4. Prepare known positive control in Gey's buffer at 10X final concentration (e.g., if using fMLP prepare a $10^{-7}M$ solution).
5 Aliquot 270 µL of cell suspension into sterile tubes.
6. Add 30 µL of chemoattractant.
7. Vortex tubes and incubate for 10 min in water bath at 37°C.
8. Add 1 mL of ice cold paraformaldehyde to fix cells.
9. Incubate for 1 h at 4°C.
10. Wash cells twice by centrifuging at 150*g* for 10 min at 4°C with 2 mL of ice-cold PBS.
11. After final spin, aspirate PBS and add approx 100–200 µL fresh PBS and resuspend cells.
12. Make a wet mount by adding approx 10–25 µL of cell suspension onto a microscope slide, and cover slip.

13. The percent polarization is determined by counting a minimum of 200 consecutive cells.
14. Cells can be left overnight in tubes without adversely affecting the final result.

3.4. Chemotaxis assays

3.4.1. Modified Boyden Chemotaxis Chamber Assay

1. Each sample, including positive and negative controls, needs to be set up in triplicate.
2. Prepare leukocytes as described above. One requires 200 µL of 2×10^6 cells/mL for each chamber.
2. Prepare appropriate dilutions of test chemoattractant. Each chamber requires approx 1.5 mL of the sample. We routinely make 5 mL of each dilution.
3. Prepare appropriate positive control (a total volume of 5 mL is required).
4. Assemble the Boyden chambers.
5. Using a pair of fine forceps carefully place the appropriate filter into the chamber (8-µm pore size for monocytes and lymphocytes; 3-µm pore size for neutrophils) by holding the filter at its very edge. Take care not to puncture filter, and not to handle filter other than with forceps.
6. We have our Boyden chambers numbered, and therefore do not number the filters. However, filters may also be numbered on the outer edge using a pencil. This avoids any confusion should filters accidently get out of order during the staining procedure and also helps to orient the top of the filter from the undersurface of the filter when mounting it on the slide.
7. Screw the upper chamber cap making sure that it is secure so as not to allow the reagents to leak. However, do not tighten the cap excessively, since this often causes the fragile filter to crack.
8. Angle the chamber at approx 35–40° (it is sometimes convenient to balance the chamber against the edge of a test tube rack).
9. Working rapidly, deliver approx 1 mL of the chemoattractant to the lower chamber, using a sterile Pasteur pipet.
10. Make sure that no air bubbles get trapped under the filter (the BSA tends to froth readily, so take care when pipetting the reagent).
11. As soon as the fluid wets the filter (the filter turns gray and translucent, as opposed to white and opaque), gradually level the chamber and deliver 200 µL of the cell suspension to the top chamber. This step needs to be performed rapidly, such that the chemoattractant does not enter the top chamber.
12. Add extra 2% BSA in Gey's buffer to the chamber with the cells, so that the fluid is level with the top ofthe cap. To the lower chamber, add the remaining chemoattractant such that levels of the fluids are at the same height on either side of the filter.
13. Incubate the chambers in a humidified incubator (30 min for neutrophils, 2 h for lymphocytes and monocytes).

14. At the end of the incubation period aspirate the fluid from either side of the filter.
15. Remove the filter from the chamber and fix the filter in absolute ethanol for 1 min.
16. Stain filters in hematoxylin (7 min).
17. Rinse in water.
18. Dip in 1% sodium acetate (5 s).
19. Rinse in water.
20. Dehydrate in 50% ethanol (30 s), 70% ethanol (1 min), 95% ethanol (1 min), absolute ethanol (1 min), absolute ethanol (1 min).
21. Clear in two lots of xylene (3 min each).
22. Place filter on slide, top surface up.
23. Add a fairly large drop of Permount and cover slip. Gently push down on the cover slip so that the Permount spreads to the edges of the cover slip. Add sufficient Permount to cover filter without introducing any air bubbles. However, care should be taken not to load on too much Permount, since it is difficult to focus through the filter if the mounting medium is too thick.
24. Chemotaxis measured by the leading front method is determined as follows. Focus on the top of the slide, and determine the reading on the micrometer gauge on the fine focus knob of the microscope. Focus through the filter until one finds the most forward moving cell. Now rack back on the focus until one finds at least two cells in the field. Take the reading on the micrometer gage. The difference between the two readings gives the distance migrated in micrometers. Repeat this procedure on the slide four more times. Calculate the average of the five readings.
25. Repeat this procedure for each of the filters.
26. We routinely read our slides "blind," and decode the slides once all determinations have been made (*see* **Notes 10–13**).
27. Slides can be kept overnight if readings cannot be made on the same day. However, it is necessary to ensure that the slides do not dry out. If the filters accidently dry out. Remove the cover slip, by soaking the slide in xylene, dip filter in xylene for a few minutes, and then remount. The filter will once again become transparent and readings can be taken.

3.4.2. Multiwell Chemotaxis Chamber Assay

1. Prepare bottom chamber of multiwell chamber by dispensing 25 μL of chemoattractant solution and control solutions to wells.
2. Using forceps, layer the filter carefully, making sure there is no crosscontamination between contents of the various wells.
3. Apply gasket and top plate.
4. Hold top plate firmly against bottom plate while attaching the thumb nuts. Do not over tighten thumb nuts. *Important:* Do not let any air bubbles get in below the filter.
5. Add 50 μL of leukocyte suspension to the top wells. Once again make absolutely sure that no air bubbles get trapped in these wells.

6. Incubate chamber in humidified air with 5% carbon dioxide for 30–45 min for neutrophil migration, and 2 h for monocyte migration.
7. Remove filters at the end of the incubation period and gently remove non-migrating cells with wiper blade as described in the manufacturer's instructions.
8. Fix membrane in methanol and stain with Diff Quik or hematoxylin as described above for modified Boyden chemotaxis chamber assay (*see* **Notes 14** and **15**).
9. Place filter, topside-up and measure distance migrated through filter, as described for modified Boyden chemotaxis chamber assay.

3.5. Chemokinesis: "Checkerboard Assay"

1. Set up a checkerboard assay as indicated in Table 1 such that there is an increasing amount of chemoattractant both below as well as above the filter.
2. Perform assay exactly according to method described above in **Subheading 3.4.1.**, using triplicates for each assay point.
3. Calculate average distance migrated according to the leading front method for each of the concentrations.
4. If one gets migration of cells in the absence of a gradient (i.e., along the diagonal) when equal concentrations of chemoattractant are placed both above and below the filter, this is a reflection of chemokinetic motion.

4. Notes

1. When using human blood always use **caution:** Gloves, masks, and lab coats should always be worn. We routinely perform our cell separations in a biosafety hood. Proper laboratory safety practices should be used when disposing of materials (centrifuge tubes, pipets, and so on) that may have come in contact with human blood.
2. To get the best results from functional assays it is ideal to have cells that are as close as possible to a resting state. Therefore, always treat cells gently. Do not use vigorous pipetting or vortexing. Always remove cells from the centrifuge as soon as it stops. Do not let cells sit for a long period of time in a pellet. Perform assays as soon as possible after cell purification has been achieved.
3. To ensure a high cell yield always centrifuge cells at room temperature when using cell separation gradients such as LSM or Nycoprep, since the specific gravity of these reagents alter with temperature.
4. Ensure that all reagents (separation media and buffers) are endotoxin free.
5. Endotoxin-free BSA is costly therefore prepare just sufficient for each experiment.
6. We follow the very comprehensive instructions provided by the manufacturer to determine the presence of endotoxin. All reagents and buffers should have endotoxin levels well below the lowest standard as provided with the kit, otherwise they could adversely affect the results obtained in the chemotaxis assays.
7. Although the manufacturer suggests pyrogen-free glass assay tubes, we have not had any problems associated with the polystyrene tubes we use, since they are used only for diluting out the standards.

8. Vigorous vortexing of standards before dispensing into the microtiter plate is essential.
9. It is important that BSA is left out of all of the buffers in this assay, since BSA has chemokinetic properties and may cause polarization of cells in the absence of the chemoattractant.
10. This is the assay of choice in our laboratory. Results are consistent and reproducible.
11. The one drawback is that it does require fairly large amount of reagents.
12. The most critical step is the rapid addition of cells to the upper chamber just as soon as the chemoattractant "wets" the filter.
13. The alternative to using the leading front method is the use of polycarbonate filters and measuring the numbers of cells migrated to the under surface of the filter. However, we have found this technique to be less reliable than the leading front technique and do not as a rule use it unless we employ the Transwell™ cell culture chambers.
14. The advantage of this technique is that it requires small amounts of reagents and large numbers of samples can be set up.
15. However, we have found that it is rather more difficult to set this apparatus up without trapping air bubbles in either the upper or lower wells.

Acknowledgment

My thanks to Ramesh Hegde for his careful readings of this manuscript and for the numerous chemotaxis assays performed with meticulous and painstaking skill. Work reported from this laboratory has been supported in part by Public Health Service grant AI-28018 from the National Institute of Allergy and Infectious Disease and grants from the American Cancer Society, the Oklahoma Center for Advancement of Science and Technology, and the Alzheimer's Association.

References

1. Pereira, H. A. (1995) CAP37, a neutrophil-derived multifunctional inflammatory mediator. *J. Leukoc. Biol.* **57,** 805–812.
2. Streiter, R. M., Kasahara, K., Allen, R. M., Standford, T. J., Rolfe, M. W., Becher, F. S., Chensue, S. W., and Kunkel, S. L. (1992) Cytokine-induced neutrophil-derived interleukin-8. *Am. J. Pathol.* **141,** 397–407.
3. Pereira, H. A., Shafer, W. M., Pohl, J., Martin, L. E., and Spitznagel, J. K. (1990) CAP37, a human neutrophil-derived chemotactic factor with monocyte specific activity. *J. Clin. Invest.* **85,** 1468–1476.
4. Territo, M. C., Ganz, T., Selsted, M. E., and Lehrer, R. I. (1989) Monocyte chemotactic activity of defensins from human neutrophils. *J. Clin. Invest.* **84,** 2017–2020.
5. Flodgaard, H., Ostergaard, E., Bayne, S., Svendsen, A., Thomsen, J., Engels, M., and Wollmer, A. (1991) Covalent structure of two novel neutrophile leucocyte-

derived proteins of porcine and human origin: neubophil elastase homologues with strong monocyte and fibroblast chemotactic activities. *Euro. J. Biochem.* **197,** 535–547.

6. Henson, P. M., Doherty, D. E., Riches, D. W. H., Parsons, P. E., and Worthen, G. S. (1994) LPS and cytokines in lung injury, in *Endotoxin and the Lungs* (K.L. Brigham, ed.), Marcel Dekker, NY, pp. 267–304.
7. Bing, D. H., Feldman, R. J., and Fenton, J. W. (1986) Structure-function relationships of thrombin based on the computer generated three-dimensional model of the β-chain of bovine thrombin. *Ann. NY Acad. Sci.* **485,** 104–119.
8. Kawaguchi, T., Ueda, K., Yamamoto, T., and Kambara, T. (1984) The chemical mediation of delayed hypersensitivity skin reactions. IV. Activation of chemotactic precursor by a trypsin-like protease in guinea pig plasma. *Am. J. Pathol.* **115,** 307–315.
9. Rose, E. M., Meromsky, L., Setter, E., Vinter, D. W., and Goldberg, I. D. (1990) Purified scatter factor stimulates epithelial and vascular endothelial migration. *Proc. Soc. Exp. Bio. Med.* **195,** 34–43.
10. Wilkinson, P. C. (1992) *Chemotaxis and Inflammation, 2nd ed.* Churchill Livingstone, Edinburgh, UK, pp. 119–142.
11. Fulk, W., Goodwin, R. H., and Leonard, E. J. (1980) A 48-well micro chemotaxis assembly for rapid and accurate measurement of leukocyte migration. *J. Immunol. Methods* **33,** 239–247.
12. Zigmond, S. H. and Hirsch, J. G. (1973) Leukocyte locomotion and chemotaxis: new methods for evaluation, and demonstration of a cell-derived chemotactic factor. *J. Exp. Med.* **137,** 387–410.
13. Kukita, I., Yamamoto, T., Kawaguchi, T., and Kambara, T. (1987) Fifth component of complement (C5)-derived high-molecular-weight macrophage chemotactic factor in normal guinea pig serum. *Inflammation* **11,** 459–479.

17

Neutralization of the In Vivo Activity of *E. Coli*-Derived Lipopolysaccharide by Cationic Peptides

Daniel J. Brackett, Megan R. Lerner, and H. Anne Pereira

1. Introduction

It has been established that at least three of the neutrophil-derived cationic antibiotic proteins, in addition to their antibiotic activity, can bind lipopolysaccharide. It is generally accepted that the binding to lipopolysaccharide, an outer membrane component of Gram-negative bacteria, is a necessary step for the antibacterial activity of CAP37 (cationic antimicrobial protein of 37 kDa molecular weight) ***(1,2)***, bactericidal permeability increasing protein (BPI/CAP57) ***(3)***, and CAP18 ***(4)***. It is apparent that a substance possessing the capacity to bind lipopolysaccharide and neutralize its biological activity has strong potential for use as a therapeutic intervention in sepsis. However, no matter how convincing the in vitro data, these substances and their derivatives must be tested in in vivo model systems for full characterization of binding and neutralization potency. Lipopolysaccharide-binding and neutralization of its biologic activity can be detected and quantified by monitoring specific in vivo responses induced by intravascular introduction of lipopolysaccharide into conscious, instrumented animals. The instrumentation of these animals permits measurement of definitive hemodynamic parameters and provides access to arterial and venous blood for determination of metabolic, hormonal, and immunologic function ***(5–7)***. It is critical that these measurements are made in conscious animals since anesthetics have detrimental effects on the basal biological status as well as compensatory responses to a changing biological system such as those occurring after a lipopolysaccharide challenge ***(8–10)***.

From: *Methods in Molecular Biology, Vol. 78: Antibacterial Peptide Protocols*
Edited by: W. M. Shafer, Humana Press Inc., Totowa, NJ

When lipopolysaccharide is introduced intravascularly, cardiovascular responses begin rapidly with the direction and magnitude of the response dependent upon the concentration and duration of delivery time ***(5–11)***. Infusion of lipopolysaccharide at low concentrations produces increased cardiac output and decreased systemic vascular resistance, which is a *hyperdynamic* cardiovascular state, whereas rapid administration of higher concentrations produce decreased cardiac output accompanied by increased systemic vascular resistance, a cardiovascular state termed *hypodynamic*. If sustained, either of these hemodynamic conditions result in compromised tissue perfusion and subsequent metabolic dysfunction and eventually tissue injury and cellular death. It is also necessary to monitor cardiac output, which allows determination of systemic vascular resistance and the opportunity to evaluate total systemic blood flow and tissue perfusion. This significant evaluation cannot be derived from blood pressure alone since blood pressure is simply a product of cardiac output (flow) and the contractile status of the vascular system (resistance); therefore, a normal blood pressure can occur during either a hyperdynamic or hypodynamic cardiovascular state and compromised tissue perfusion.

The lipopolysaccharide-induced impairment of tissue perfusion indicated by the cardiovascular responses can be substantiated by significantly increased blood concentrations of lactate providing evidence of a shift from aerobic to anaerobic metabolism eventuating in cellular injury and lethality. Additionally, lipopolysaccharide induces the synthesis of tumor necrosis factor resulting in elevated circulating concentrations and evokes the expression of endothelial cell and leukocyte adhesion molecules and their ligands with the subsequent increase in the marginating leukocyte pool and decrease in the circulating pool. The ability to document comprehensive biologic responses to lipopolysaccharide provides a panel of in vivo assays to determine the extent of the neutralization potential of antibacterial peptides. These significant cardiovascular and metabolic alterations, leukocyte and cytokine responses, and toxicity induced by lipopolysaccharide are neutralized by a 2.7-kDa cationic peptide, corresponding to residues 20–44 of CAP37 ***(12,13)***, when administered simultaneously with lipopolysaccharide or after the lipopolysaccharide has been introduced intravascularly. These in vivo data confirm the potent lipopolysaccharide-neutralization properties found in in vitro studies demonstrating that this peptide (CAP37 $P_{20–44}$) is bactericidal for *E. coli* ***(13)*** and also attenuates tumor necrosis factor production from lipopolysaccharidestimulated monocytes and neutralizes lipopolysaccharide reactions in the *Limulus* amoebocyte lysate assay.

Peptide or antibody binding to lipid A, the biologically active component of lipopolysaccharide, does not necessarily neutralize the in vitro or the in vivo activity of the molecule ***(14,15)***; and there is not an obligatory association between neutralization of in vitro interactions and in vivo biologic activity ***(14,16)***. These differences in responses may be owing to distinct and separate lipid A epitopes or substructures that elicit different responses ***(17,18)*** and provide discrete binding sites for diverse antilipopolysaccharide reagents ***(19,20)***. Indeed, the dynamics of binding and neutralization that occur in vitro do not appear to be analogous to those occurring in vivo ***(21)***. Ongoing research in this area has elucidated the complexity of the relationship between the structure and function of the lipopolysaccharide molecule and has emphasized the necessity of comprehensive in vivo, as well as in vitro testing, such as has been demonstrated with CAP37 P_{20-44}, if definitive characterization of the neutralization potential of a binding peptide is desired. These models provide the opportunity to assay an additional array of lipopolysaccharide-elicited in vivo responses to determine the neutralization capacity of the peptide of interest. These responses include increased hematocrit (intravascular fluid loss), decreased pH and altered blood gases (metabolic acidosis), increased blood concentrations of vasoactive hormones (e.g., catecholamines, vasopressin, histamine) ***(5,6)***, and cytokines (IL-1, -6, and -8). Importantly, tissue can be collected for histologic evaluation of cellular damage and determination of excessive generation of mediators of tissue injury (e.g., reactive oxygen species) evoked by lipopolysaccharide ***(22–25)***.

2. Materials

2.1. Evaluafion of Biologic Responses

2.1.1. Hemodynamic (Hyperdynamic and Hypodynamic)

1. Male Sprague-Dawley rats: 300 ± 25 g.
2. Infant laryngoscope with modified Miller 0 blade.
3. Tracheal tube: PE-240, 9.5 cm total length, 5.5 cm insertion length.
4. Anesthesia machine: isoflurane vaporizer.
5. IsoFlo®: isoflurane, USP (Abbott, North Chicago, IL).
6. Rat ventilator (Harvard, South Natick, MA).
7. Surgical instruments.
8. Arterial and venous catheters: PE and silastic tubing.
9. Thermodilution cardiac output thermistor (Columbus, Columbus, OH).
10. Lidocaine hydrochloride, 10 mg/mL (Abbott).
11. Monitoring chambers with catheter swivels (Columbus).
12. Pressure transducers (Gould Statham, Cleveland, OH).
13. Thermodilution cardiac output computer with chart recorder (Columbus).
14. Physiograph (Gould).
15. Peptide to be assayed.
16. Lipopolysaccharide, *E. coli* 0127:B8 (Sigma, St. Louis, MO).
17. Saline: sterile, nonpyrogenic (Baxter, Deerfield, IL).

18. Heparin: sodium injection, 1000 U/mL (Upjohn, Kalamazoo, MI).
19. Infusion pumps (Razel, Stamford, CN).
20. Sodium pentobarbital: 50 mg/mL (Abbott).

2.1.2. Humoral and Cellular Parameters

1. Lactate/glucose analyzer (Yellow Springs, Yellow Springs, OH).
2. Micro-capillary centrifuge and hematocrit reader (International Equipment, Needham Hts., MA).
3. pH/Blood gas analyzer (Instrumentation Laboratories, Lexington, MA).
4. Hemacytometer (Scientific Products, Grand Prairie, TX).
5. Microscope slides: precleaned, frosted (Fisher, Plano, TX).
6. Wright Giemsa stain (Curtin Matheson, Houston, TX).
7. Slide Stainer (Ames Hema Tek, Elkhart, IN).
7. Light microscope.

2.1.3. Tissue Injury

1. Tissue cassettes: OmniSette (Fisher).
2. Buffered formalde-fresh, 10% formalin (Fisher).
3. Alcohol, 70%.
4. Light microscope.

2.2. Assessment of Toxicity

1–7. Same as **Subheading 2.1.1.**
8. Venous catheter: PE tubing.
9. Xylocaine.
10. Actinomycin D (Calbiochem-Novabiochem, LaJolla, CA).
11. Lipopolysaccharide, *E. coli* 0127:B8.
12. Peptide to be assayed.
13. Sodium pentobarbital.

3. Methods

3.1. In Vivo Instrumentation Procedures

3.1.1. Evaluation of Biological Responses

1. Induce anesthesia using a bell jar containing the rapidly metabolized, inhalational anesthetic, isoflurane (5%) (*see* **Note 1**).
2. Intubate the trachea.
3. Connect the tracheal tube to a rodent respirator delivering 2.5% isoflurane combined with 100% O_2 delivered at 600 mL/min.
4. Make a 2-cm midline incision through the skin on the ventral side of the neck.
5. Isolate the right carotid artery and jugular vein.
6. Insert a venous catheter with the tip at the level of the right heart for measurement of central venous pressure, blood sampling, and infusion of saline, lipopolysaccharide, and/or the peptide to be evaluated.

7. Via the artery, insert a thermocouple catheter (described in **ref. *1***) so the tip is placed immediately distal to the aortic valve for measurement of arterial blood pressure, blood sampling, and generation of thermodilution cardiac output curves.
8. Guide the catheters under the skin to exit through an incision in the back of the neck just below the base of the skull.
9. Inject skin edges of the incision with lidocaine.
10. Suture wounds and secure catheters at the exit incision using 4–0 surgical silk.
11. Remove the animal from the respirator and delivery of isoflurane.
12. Upon resumption of normal respiration and demonstration of the righting reflex remove the intratracheal tube.
13. Using the swivel in the monitoring chamber connect the catheters to the pressure transducers, the physiograph, and the cardiac output computer.
14. Maintain close observation for the remainder of the experimental protocol testing for acute responses (*see* **Note 2**).

3.1.2. Assessment of Toxicity

1. Repeat **steps 1–12** in **Subheading 3.1.1.**, with the exception that only the venous catheter is inserted.
2. Return animals to home cages with free access to food and water for 48 h before testing the efficacy of the chosen peptide on long-term LPS-induced lethality.

3.2. Biological Responses

3.2.1. Hemodynamic

1. Allow a 60-min recovery period beginning at cessation of the anesthetic.
2. Monitor cardiovascular parameters for 30 min to assure stable, normal readings and collect control data prior to lipopolysaccharide (LPS).
3. Administer LPS at the appropriate concentration to produce the desired response:
    a. Hyperdynamic response (high cardiac output/low systemic vascular resistance); 30 min infusion of LPS at a concentration of 250 µg/kg in 1 mL of sterile saline.
    b. Hypodynamic response (low cardiac output/high systemic vascular resistance); intravenous LPS bolus at a concentration of 15–20 mg/kg/mL.
4. Administer the peptide at concentrations to be evaluated at the appropriate time to address specific issues regarding the effectiveness of LPS binding:
    a. Peptide preincubation with LPS.
    b. Peptide delivered simultaneously with LPS.
    c. Peptide delivered at selected times after initiation of LPS administration.
5. Record cardiovascular responses (arterial blood pressure, cardiac output, heart rate, and central venous pressure) 5, 15, 30, 45, 60, 90, 120, 180, and 240 min after the initiation of LPS administration (*see* **Note 5**).
6. Calculate systemic vascular resistance [(mean arterial pressure – central venous pressure)/cardiac output] and cardiac stroke volume (cardiac output/heart rate) for each of the time-points.

7. Induce euthanasia using intravenous sodium pentobarbital (45 mg/kg).
8. Confirm that the thermistor and venous catheter tips are properly oriented (*see* **Note 4**).

3.2.2. Humoral and Cellular

1. During the protocol described above collect arterial blood samples (200 µL) before LPS administration and at 60 and 240 min after LPS.
2. Replace each blood withdrawal with an equal volume of sterile saline.
3. Fill catheter with heparinized saline at the exact, predetermined volume of the catheter (*see* **Note 3**).
4. Divide each blood sample into appropriate aliquots for measurement of blood lactate and glucose concentrations (25 µL), hematocrits (70 µL), pH and blood gases (65 µL), and total white blood cell and differential counts (40 µL).
5. Obtain values for each parameter using the equipment described above in **Subheading 2.1.2.**
6. If measurement of blood concentrations of cytokines, vasoactive substances, enzymes, or other mediators known to be induced by LPS are necessary to evaluate the peptide, the amount of blood required by the appropriate assay can be collected during these protocols.

3.2.3. Tissue Injury

1. Following euthanasia, collect tissue samples from organs of interest: small intestine, liver, kidney, lungs.
2. Place tissue samples in tissue cassettes.
3. Place cassettes in formalin.
4. After 24 h, place cassettes in alcohol until processed for histology (*see* **Note 6**).

3.3. Assessment of Toxicity

1. Forty-eight hours after implantation of venous catheters as described in **Subheading 3.1.**, administer an iv bolus infusion of the combination of actinomycin D (800 µg/mL/kg) and lipopolysaccharide (2.5 µg/mL/kg) in sterile saline.
2. Following the lethal lipopolysaccharide (LPS) challenge deliver an iv bolus of the desired concentration of peptide, control peptide, or carrier solution.
3. The time of delivery of the peptide or the associated controls relative to the administration of LPS should be determined by the hypothesis to be tested involving the efficacy of the peptide.
4. Record lethality for 7 d to collect data for comparison of survival times and permanent survival rates.

3.4. Data Analysis

1. Groups should contain a minimum of 10 animals.
2. Repeated measures analysis of variance (ANOVA) should be used to evaluate the hemodynamic, respiratory, and humoral parameters.

3. If significant main or interactive effects are found from the ANOVA test ($p \leq 0.05$), simple contrasts between-groups or within-group (relative to time zero) should be made using the appropriate unpaired or paired *t*-test or Duncan's new multiple range test (*post hoc*).
4. The Chi-square test should be applied to test for differences in lethality between groups.

4. Notes

1. Use rapidly metabolized inhalational anesthesia.
2. Study protocols utilizing conscious animals require emphasis on a low-key, quiet, temperature controlled environment.
3. Catheters must be filled with heparinized saline to prevent clotting, but care must be taken not to overfill and heparinize the animal.
4. It is imperative to confirm correct placement of the thermistor tip (immediately distal to the aortic valve) and the tip of the venous catheter (adjacent to the right atrium) at the termination of each study to assure proper generation of cardiac output curves.
5. A glass 100-μL syringe (Hamilton) should be used for injection of saline for generation of thermodilution curves.
6. Evaluation of histology should be done in a coded fashion by a pathologist experienced with rat tissue morphology.

Acknowledgments

We wish to acknowledge the financial support of the Department of Surgery Research Funds, University of Oklahoma Health Sciences Center; American Cancer Society, # IM 71966; Presbyterian Health Foundation; National Institutes of Health, # AI 28018; and Department of Veterans Affairs Medical Research Service.

References

1. Shafer, W. M., Martin, L. E., and Spitznagel, J. K. (1984) Cationic antimicrobial proteins isolated from human neutrophil granulocytes in the presence of diisopropyl fluorophosphate. *Infect. Immun.* **45,** 29–35.
2. Shafer, W. M., Martin L. E., and Spitznagel J. K. (1986) Late intraphagosomal hydrogen ion concentration favours the in vitro antimicrobial capacity of a 37 kD cationic granule protein of human neutrophil granules. *Infect. Immun.* **53,** 651–655.
3. Marra, M. N., Wilde, C. G., Collins, M. S., Snable, J. L., Thornton, M. B., and Scott, R. W. (1992) The role of bactericidal/permeability-increasing protein as a natural inhibitor of bacterial endotoxin. *J. Immunol.* **148,** 532–537.
4. Larrick, J. W., Hirata, M., Shimokmoura, Y., Yoshida, M., Zhong, J., and Wright, S. C. (1993) Antimicrobial activity of rabbit CAP18-derived peptides. *Antimicrob. Agents Chemother.* **37,** 2534–2539.

5. Brackett, D. J., Schaefer, C. F., Tompkins, P., Fagraeus., L., Peters, L. J., and Wilson, M. F. (1985) Evaluation of cardiac output, total peripheral vascular resistance, and plasma concentrations of vasopressin in the conscious, unrestrained rat during endotoxemia. *Circ. Shock* **7,** 273–284.
6. Brackett, D. J., Hamburger, S. A., Lerner, M. R., Jones, S. B., Schaefer, C. F., Henry, D. P., and Wilson, M. F. (1990) An assessment of plasma histamine concentrations during documented endotoxic shock. *Agents Actions* **31,** 263–274.
7. Brackett, D. J., Gauvin, D. V., Lerner, M. R., Holloway, F. A., and Wilson, M. F. (1994) Dose- and time-dependent cardiovascular responses induced by ethanol. *J. Pharmacol. Exper. Therapeut.* **268,** 78–84.
8. Shaefer, C. F., Brackett, D. J, Tompkins, P., and Wilson, M. F. (1984) Anesthetic-induced changes in cardiovascular and small intestinal responses to endotoxin in the rat. *Adv. Shock Res.* **10,** 125–133.
9. Biber, B., Schaefer, C. F., Smolik M. J., Lawrence, M. C., Lerner, M. R., Brackett D. J., Wilson, M. F., and Fagraeus, L. (1987) Dose-related effects of isoflurane on superior mesenteric vasoconstriction induced by endotoxemia in the rat. *Acta Anaesthesiol. Scand.* **31,** 430–447.
10. Schaefer, C. F., Biber, B., Brackett, D. J., Schmidt C. C., Fagraeus, L., and Wilson, M. F. (1987) Choice of anesthetic alters the circulatory shock pattern as gauged by conscious rat endotoxemia. *Acta Anaesthesiol. Scand.* **31,** 550–556.
11. Brackett, D. J., Lerner, M. R., and Pereira, H. A. (1994) An LPS-binding peptide derived from the neutrophil-associated protein CAP37 prevents endotoxin induced hyper- and hypo-dynamic responses in the conscious rat. *J. Endotoxin Res.* **1,** 93.
12. Pohl, J., Pereira, H. A., Martin, N. M., and Spitznagel, J. K. (1990) Amino acid sequence of CAP37, a human neutrophil granule-derived antibacterial and monocyte-specific chemotactic glycoprotein structurally similar to neutrophil elastase. *FEBS Lett.* **272,** 200–204.
13. Pereira, H. A., Erdem, I., Pohl, J., and Spitznagel, J. K. (1993) Synthetic bactericidal peptide based on CAP37: a 37-kDa human neutrophil granule-associated cationic antimicrobial protein. *Proc. Natl. Acad. Sci. USA* **90,** 4733–4737.
14. Parent, J. B., Gazzano-Santoro, H., Wood, D. M., Lim, E., Pruyne, P. T., Trown, P. W., and Conion, P. J. (1992) Reactivity of monoclonal antibody E5 with endotoxin. II. Binding to short- and long-chain smooth lipopolysaccharides. *Circ. Shock.* **38,** 63–73.
15. Warren, H. S., Amato, S. F., Fitting, C., Black, K. M., Loiselle, P. M., Pasternack, M. S., and Cavaillon, J.-M. (1993) Assessment of ability of murine and human anti-lipid A monoclonal antibodies to bind and neutralize lipopolysaccharide. *J. Exp. Med.* **177,** 89–97.
16. Gudmundsson, S. and Craig, W. A. (1986) Role of antibiotics in endotoxin shock, in *Handbook of Endotoxin* (Proctor, R. A., ed.), Elsevier, Amsterdam, pp. 238–263.
17. Lasfargues, A., Tahri-Jouti, M.-A., Pedron, T., Girard, R., and Chaby, R. (1989) Effects of lipopolysaccharide on macrophages analyzed with anti-lipid A monoclonal antibodies and polymyxin B. *Eur. J. Immunol.* **19,** 2219–2225.

18. Tahri-Jouti, M.-A., Mondange, M., Le Dur, A., Auzanneau, F. I., Charon, D., Girard, R., and Chaby, R. (1990) Specific binding of lipopolysaccharides to mouse macrophages. II. Involvement of distinct lipid A substructures. *Molec. Immunol.* **27,** 763–770.
19. Fujihara, Y., Bogard, W. C., Lei, M.-G., Daddona P. E., and Morrison, D. C. (1993) Monoclonal anti-lipid A IgM antibodies HA-1A and E-5 recognize distinct epitopes on lipopolysaccharide and lipid A. *J. Infect. Dis.* **168,** 1429–1435.
20. Pedron, T., Girard, R., Lasfargues, A., and Chaby, R. (1993) Differential effects of a monoclonal antibody recognizing 3-deoxy-D-manno-2-octulosonic acid on endotoxin induced activation of pre-B cells and macrophages. *Cell. Immunol.* **148,** 18–31.
21. Arden, W. A., Strodel, W. E., Gross, D. R., Anderson, K. W., Oremus, R., Derbin, M., and Schwartz, R. W. (1995) Preincubation of endotoxin with monoclonal anti-lipid A (E5), but not in vivo treatment, inhibits circulatory dysfunction. *Shock* **4,** 131–138.
22. Brackett, D. J., Lai, E. D., Lerner, M. R., Wilson, M. F., and McCay, P. B. (1989) Spin trapping of free radicals produced in vivo in heart and liver during endotoxemia. *Free Rad. Res. Commun.* **7,** 315–324.
23. Brackett, D. J., Lerner, M. R., Wilson, M. F., and McCay, P. B. (1995) Evaluation of in vivo free radical activity during endotoxin shock using scavengers, electron microscopy, spin traps, and electron paramagnetic resonance spectroscopy. *Adv. Exp. Med. Biol.* **366,** 407–409.
24. Balla, A. K., Doi, E. M., Lerner, M. R., Bales, W. D., Archer, L. T., Wunder, P. R., Wilson, M. F., and Brackett, D. J. (1996) Dose-response effect of in vivo endotoxin on polymorphonuclear leukocytes (PMN's) Oxidative Burst. *Shock* **5,** 357–361.
25. Wallis, G., Brackett, D. J., Lerner, M. R., Kotake, Y., Bolli, R., and McCay, P. B. (1996) In vivo spin trapping of nitric oxide generated in the small intestine, liver, and kidney during the development of endotoxemia. *Shock* **6,** 274–278.

Index